Effective Environmental Emergency Responses

Paul A. Erickson

Effective Environmental Emergency Responses

A Holistic Response

Springer

Paul A. Erickson
Anna Maria College
Paxton, MA, USA

ISBN 978-3-031-05892-9 ISBN 978-3-031-05893-6 (eBook)
https://doi.org/10.1007/978-3-031-05893-6

This Springer imprint is published by the registered company Springer Nature Switzerland AG
The registered company address is: Gewerbestrasse 11, 6330 Cham, Switzerland

For all those I have been privileged to call "my students," including undergraduates, graduates, and practicing professionals, as well as those I forever revere as my mentors, and for Christine Holmes, my love, partner, and guiding spirit.

Preface

A fundamental biological truism is that all forms of life are daily faced with threats to their survival. Those that manage to survive do so either by fortuitous chance or, if they have the means, by learning new behavior and/or by changing their environment.

No matter, new threats constantly arise.

For the past couple of hundred thousand years, humans have managed to survive as a species by their cleverness despite newly encountered threats and the perpetual loss of individuals. More than several million years ago, early human ancestors obviously did the same; if not, we humans would not exist today. However, this similarity between relatively modern humans and our earliest ancestors does not obscure the fact that our ancestors had to overcome threats presented by nature whereas the threats we face increasingly derive from the interplay between both nature and human behavior. Thus, the subtitle of this volume, "A Holistic Perspective."

My intent is simply to explore the relevance of this perspective to a twenty-first-century approach to emergency planning and response as opposed to approaches handed down from previous centuries.

My motivation is primarily driven by my conviction that how we respond to any emergency is an appropriate measure of the human willingness to care for others. Willingness, however, is not a human instinct; we are born with only sucking, grabbing, and several muscular-reflex instincts; everything else is learned behavior.

My fondest hope is that this volume will be of practical use to undergraduate and beginning graduate students interested in the constantly growing variety of emergency-related professions.

Paxton, MA, USA Paul A. Erickson

Contents

Part I
Overview

1.1 Preface to Part I

The following 3 chapters are essentially summarized in the following concepts and ideas which comprise the basis of this entire volume. Examples and discussions in these chapters are by way of introductory examples that are more thoroughly examined in Part II.

- Scope of planning emergency response in the twenty-first century
- Emergencies as composed of interconnected causal chains, including precursors to emergencies and subsequent short- and long-term consequences
- Emergencies as exercises in the application of multi- and interdisciplinary expertise
- Expertise as applied to precursors and subsequent consequences involve all decision-making components of society, including governmental (at all levels), corporate and public institutions as well as the supportive (or non-supportive) general public.
- Wherever and whenever an emergency occurs, the planning and execution of response is always in the context of various social inequities that pervade society.
- Learning from previous failures in both planning and response must be thoroughly enmeshed in planning and response efforts. The most modern tested-technologies must be utilized in addition to historic approaches that have proven to be effective.
- The ethos of "environmental impact assessment" is a realistic tool for identifying potential precursors and ameliorating possible subsequent consequences.
- No institution is the sole possessor of expertise. A major problem is associated with expertise is how its derivative technologies are effectively communicated (or not) and how they are effectively used (or not).

Chapter 1
Introduction

First comes the right word...next comes the Right Reaction......

However the word "emergency" may be used, perhaps as a synonym for "disaster", "catastrophe", "cataclysm" (each often connoting an escalating sense of harm and/ or expanded extent of that harm), it typically implies "immediacy", which meaning is clearly explicit in the common phrase "emergency response". Of course there should be an immediate response to emergency event, but the larger question is "is the immediacy of response' enough?" Immediate response has to occur right-now.... but isn't generally viewed as extending over months, years, and even decades.

Put the fire out! And when the fire is out, the emergency and emergency response are over. No, they're not, even though the bells, whistles and sirens of warning are quiet.

Rescue the roof-top survivors of the flood! And after all are rescued, the emergency and the emergency response are over. No, they're not, yet we all too soon revert to "normal".

Over the past several decades, we have learned a great deal about the complexities of any emergency. Consider Three Mile Island Meltdown(1979), Chernobyl Meltdown (1986), Oklahoma City Bombing (1995), Hurricane Katrina (2005), Fukushima Tidal Wave (2011), California Wildfire (2020).... the dates above indicate the date of call for first-response...... but the subsequent consequences of each singular event persist to this date, and will persist for perhaps generations. We often forget (or otherwise ignore) subsequent consequences of a singular event. Who responds to the subsequent consequences? Anybody? Nobody? And what of preceding events that contributed to a specific emergency?

Thus the subtitle of this book: Identifying *& Managing Interconnected Causal Chains*. The World Health Organization (WHO) uses the term "cascade" to describe what this author describes as "interconnected causal chains", but I trust the reader will forgive my translation of "cascade" as an often unforeseen and indeterminant flow of anything: we have experience enough with emergencies to understand rather precisely likely and possible interconnections between causes and effects and,

P. A. Erickson, *Effective Environmental Emergency Responses*,
https://doi.org/10.1007/978-3-031-05893-6_1

therefore, to make sensible predictions and take appropriate ameliorative actions. Of course, the unexpected does happen. Of course we have to learn new things! We call that learning "experience".

History defines previous experience; the future depends on new knowledge ……

Any emergency typically involves a series of sequential events that also qualify as emergencies to a wide diversity of community-components, including human and non-human species as well as natural resources, environmental systems, economic structures, political systems, cultural resources…. In short, the sum-total of the components and dynamics of any social or natural system.

Because any emergency has subsequent consequences, so does emergency response require subsequent responses, each of which likely has its own time-frame, from initiation on through final resolution.

Consider, for example, the health consequences of a nuclear meltdown. They vary from immediate burn-exposure to the development and treatment of cancer; from a mutation of a fetus through health affects in that baby as well as in that baby's subsequent progeny. Does that qualify as an emergency? It certainly does to any and all individuals within that germ-line.

Or, consider that, during the Covid-19 pandemic, the volume of air-traffic was drastically reduced which, because a large amount of meteorological data is collected by commercial flights, significantly reduces our capacity to make accurate weather predictions, including such events as tornadoes and hurricanes, which subsequently delays evacuation orders for unsuspecting citizens. What's the response? As the saying goes, "who would have thought?"

In the past 40 years, the academic community has paid increasing attention to the study of emergencies. From the mid 1960's through the early 1980's, much primary attention was given to the loss of life and economic consequences of disaster. Morbidity, mortality, and economic data were easily available per-geographic location and the obvious hazards associated with floods, earthquakes and other specific hazards. By the mid 1980's, academics turned their attention also to the mapping of public attitudes and perceptions of emergencies. This approach unerringly led to the serious consideration (in the late 1990's) to risk-reduction and hazard-mitigation. It was not until the early twenty-first century that these previous considerations coalesced more and more into the multi- and interdisciplinary analysis of direct and indirect physical, chemical, biological, social, political and psychological impacts of emergencies and the mitigation of those impacts.

Effective response depends on teams of diverse sciences and skilled practitioners ……

By multi-and interdisciplinary academics, I mean the sciences, whether they be the natural or social sciences. Oh, I well know some sciences may be considered somewhat (or a whole lot) lesser than others with regard to rigor…. but no science is simply opinion or uninformed hearsay … every science is open to peer-challenge, new discovery and knowledge, even if often-times grudgingly.

The use of multi- and interdisciplinary knowledge to the planning and implementation of emergency response presents a number of difficulties with regard to the coordination of scholastic understanding and respondent practice. It is very clear that professional perspectives of both often conflict, especially given their joint participation in complex socio-political contexts. I do not hesitate to lay onus on both; they speak different languages, are evaluated by different measures, and they also demonstrate their skills in different public menus. I will discuss this in much greater detail in the three chapters of Part III of this volume. For now, suffice it to say that both scientific knowledge and practical experience are absolutely necessary for any comprehensive approach to modern emergency response planning and the execution of response.

Historically, much attention was given (and rightfully continues to be given) to emergency planning that focused on various phases of ***the emergency event ....*** from its recognized **potential** to successive stages, including (in turn), actual **early warning, crisis and response** and, ultimately (though typically imaginative) **return to normal** (Fig. 1.1).

This perspective of the genesis and supposed final demise of any emergency has proven of significant value in the planning for emergency response and the

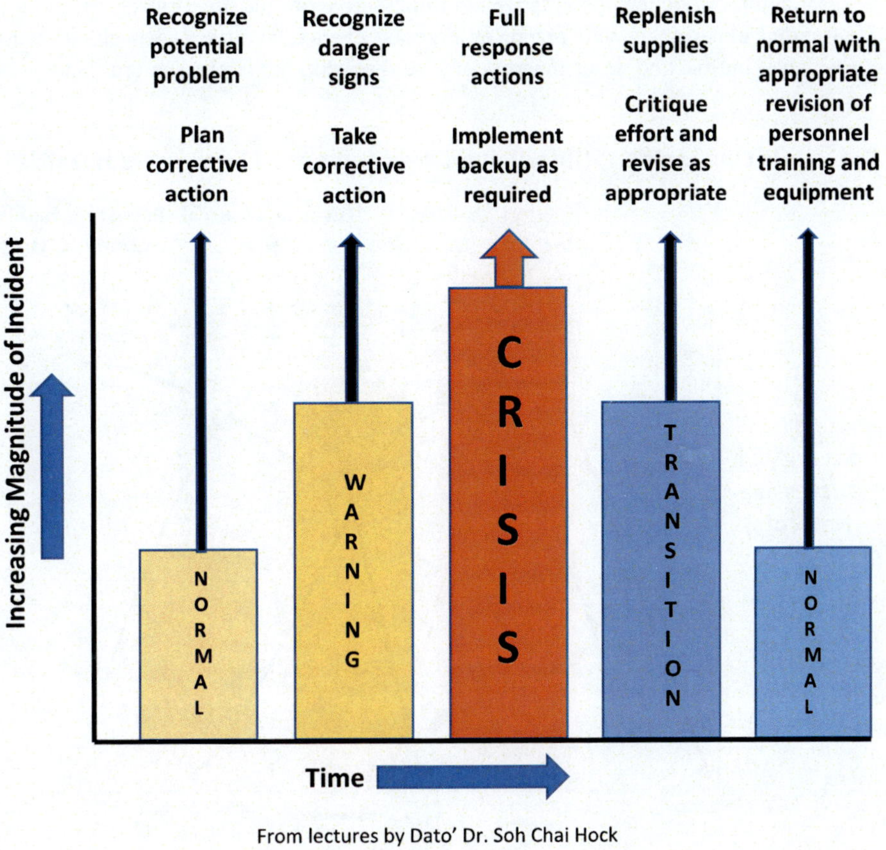

Fig 1.1 Traditional phases of the emergency event

evaluation of that response. However, it is time to expand that algorithm on the basis of our past experience to account for what we have actually experienced …… and that is simply the fact that an emergency event of any kind has a plethora of consequences that persist beyond (in time) the primary emergency event and typically coalesce to define a **New Normal**.

> Everything in this world is inter-connected, with negative and even positive consequences ……

Figure 1.2 is a very simple conceptual view of a more realistic evolution of any emergency event……a series of successive consequences of the primary (in time) emergency event. It should be emphasized that the depicted sequence of events involves conceptually very different types of phenomena *over different periods of time*, ranging from (for example) a *primary hydrological emergency* through subsequent *wildlife migrations* (of perhaps carnivores and insects) into populated areas and thereby causing in (*turn physical injury and/or disease*), which in their turn lead to *human morbidity and mortality* that exacerbate already severe familial *economic (as well as psychological) stress*.

Such inter-connected events are not imaginative prognostications; they are based on actual experience with recent emergency events. They are also elucidated by intensive multi- and inter-disciplinary studies conducted by natural and social

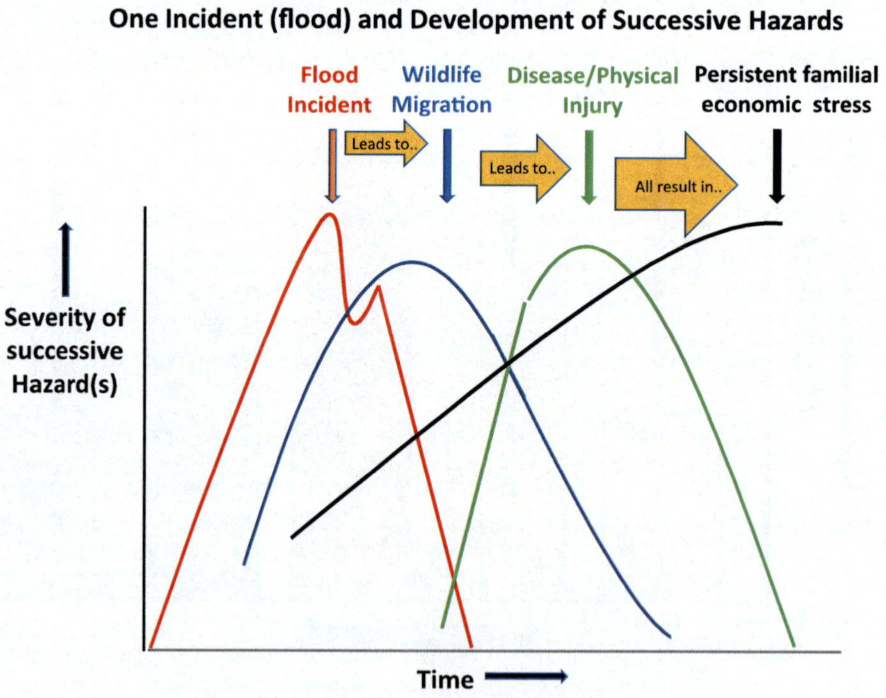

Fig. 1.2 One incident (flood) and development of successive hazards

scientists. Again, I emphasize that the adequacy of emergency response depends on both practical experience and the progress of scientific understanding which, in turn, is the basis of technology and its application.

But, of course, neither scientific knowledge nor practical experience can be applied in a vacuum; both depend in very large measure on the social-political context in which they are embedded.

> How then to build a comprehensive plan when the execution of that plan is supposedly intended to resolve an inter-connected and complex sequence of events that, one way or another, disrupt and confuse the *status quo*?

To say that the United States is diverse is utterly simplistic. It is diverse in the extreme regardless of the measure, be it in personal attitudes, biases and prejudices, income and allegiances, political preferences, religion, geographic features or social alliances. Its political and judicial structures are essentially intertwined mazes of often conflicting authority; the country's "voice", if it has one, is perhaps best described as a rather persistent cacophony (Fig. 1.3).

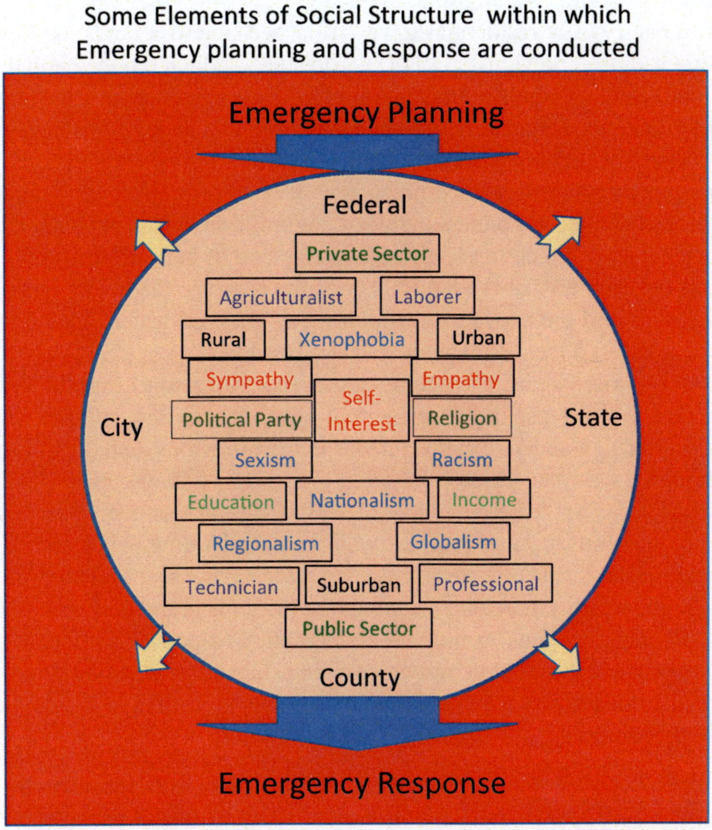

Fig. 1.3 Some elements of social structure within which emergency planning and response are conducted

The planning and responses of any emergency therefore require the cooperative efforts of diverse governmental agencies (in some instances, perhaps international teams), and as well multiple private-sectors, individuals and groups of experts and experienced practitioners, all whom must be closely coordinated within a highly complex exercise in practical logistics. Of course, such a coordination of planning and response effort is a daunting undertaking that, given even a modicum of assured success, is often (and easily) thwarted by chance and the collage of personal and regional differences with respect to public support regarding costs and perceived limits to social acceptability. Part III of this volume discusses some of these matters in more detail. Part II, however, first focuses on the other equally prime aspect of dealing with complexity, which is the interdisciplinary aspect of responding to preceding and subsequent and interconnected causal chains associated with any emergency event.

Two examples are illustrative of the importance of considering both pre- and post- emergency related and interconnected events.

Example 1.1: Highland Tower Collapse, December 11, 1993 in Taman Hillview, Ulu Klang, Selangor, Malaysia

In December 1993, a residential tower collapsed due to a landslide down a steep slope that had been previously (1974) deforested and then terraced. On the morning of the landslide, a large birthday celebration was scheduled for that tower in the early morning but, fortunately, was canceled due to a heavy rainstorm. Otherwise, many children would have been crushed. As it was, 48 residents did die.

Emergency first-responders efficiently evacuated residents of two other towers in the immediate area. Responders included not only Malaysian personnel, but also an international team which later helped to locate bodies beneath the rubble.

It was determined that the landslide was the result of the heavy rain percolating denuded soils and the consequent rupturing of clogged water pipes that had been installed to divert a local stream away from the tower.

It is appropriate as well as interesting (if not surprising) that the "emergency" was, in fact, the collapse of the residential tower. It was that collapse which set off the alarms and demanded response; it was not the previous denuding of the hillside, even though it was that which resulted in the tower's collapse even if it didn't warrant headlines as an "emergency". Nor was it the various decisions regarding the collage of decisions that had been erstwhile made regarding the location of that residential tower downside of a steep (and, unfortunately, erosion-prone) hill.

Again, the word "emergency" denotes the need for immediate action. Is that sufficient? Perhaps it's time to redefine "emergency response" to denote not only a singular spectacular event but, rather, only one slice of an on-going, time-dependent process involving how we make (or not make) decisions in light of known possibilities and probabilities.

Example 1.2: California Landslide, Highway 1, Big Sur, January 28, 2021

California is well known for its wildfires. In 2020, a wildfire enveloped over 120,000 acres, leaving denuded soils which, due to a sudden rainfall of 16 inches, became a mud wall that brushed away 150 feet of highway 1. Fortunately, no vehicles were caught in the collapse, nor travelers' lives lost. The landslide nevertheless called for immediate response to ensure safe closure of that section of highway; by definition, that was "the emergency requiring immediate response".

But, perhaps it might be more reasonable to think that the significant emergency was the previous wildfire's instigation of soil erosion in the upland ringing the highway. In both the Malaysian landslide (Example 1.1, above) and this example, it was heavy rainfall on denuded, upland soils that we well know typically results in Landslides.

Highway 1 opened in 1937 and extends through the state for over 600 miles, with particularly scenic views along Big Sur. However, given its steep slopes, this area has always demanded almost constant upkeep at very high costs.... as though there's an equal balance between someone's aesthetic pleasure and someone's heightened risk.

Again.... a fair and reasonable question: "Isn't it time to give much more serious and effectual attention to how we make (or not make) decisions in light of known possibilities and probabilities?"

Yes, the Past is prologue; however, the Present is, moment to moment, becoming the Future.

Emergencies do not just happen; they do not occur just because of Nature's or human whims but by both in often unexpected ways. Any current emergency has a constellation of contributing causes as well as subsequent effects. Successful emergency planning and response must be based on what did happen previously and on what we might guess will happen in the future. This can only be achieved with our understanding that any emergency is not simply a singular event worth headline-ink or broadcast-time but, rather, a culmination of linked actions, events and attitudes of the past and the always evolving future (Fig. 1.4). Yes, it is all very complicated. So is space-flight, but we do it.

Consider the following few facts:

1. Busy roadways have long been recognized as causing significant pollutants to their immediate surroundings. People who live in the proximity of those roadways breathe in a variety of toxic fumes and particles. Those same people are typically lower-income earners and/or marginalized populations (for whatever reason). Well, how are the routes of those roadways determined? Historically, by such issues as the ease of obtaining Rights-of-Way, local politics, economics or, depending on topography of the proposed route, relative amounts of cut-and fill. Just who (if anyone) is considering the long-range health effects on nearby residents ...... on their babies?
2. Highly processed foods are known to pose a significant increase in cancer. They also provide billions of dollars to the economy. These foods are also cheap; therefore, the risk of cancer is causally distributed among (you guessed it) lower-income (or, again, marginalized) populations.
3. Many thousands of small businesses in the U.S. have closed during the COVID 19 pandemic. The exact number is impossible to determine, largely because many do not file for bankruptcy. However, firms with fewer than 500 employees

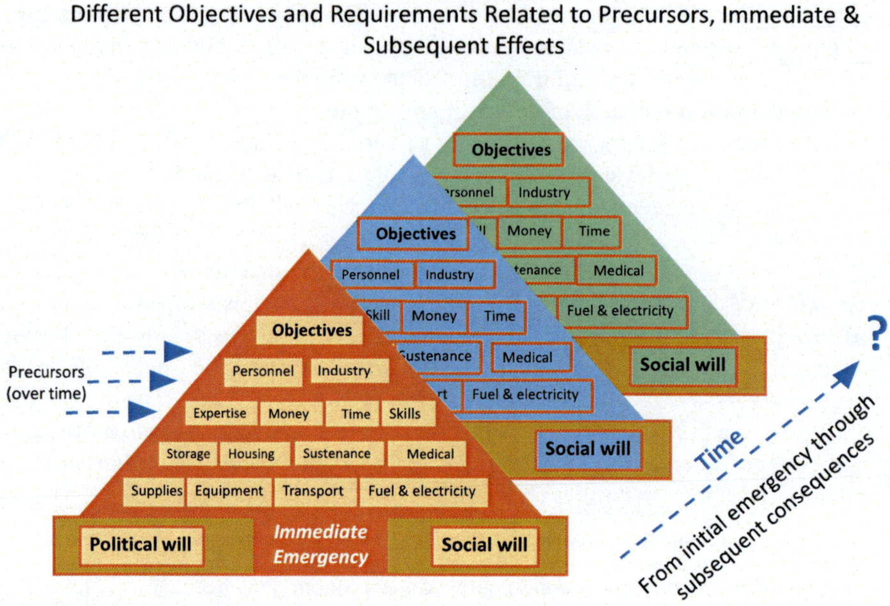

Fig. 1.4 Different objectives and requirements related to precursors, immediate & subsequent effects

> account for almost half of the U.S. economy. About half of small business-owners think they may have to permanently close. How do these owners and their employees meanwhile pay for their mortgage or rent…for their health-care, food and electricity, or other vital, day-to-day expenses?

These are only three examples of what we today commonly call "inequities"…… the unfair distribution of risks (e.g., economic, health, etc.) or social benefits to certain populations. Such social inequities have as long a history as civilization itself, but have only recently received both National and world-wide attention. Certainly, the Covid 19 pandemic has provided a critical impetus to this development; however, the global mass-media coverage of the failures as well as the successes of emergency management have also played an important role.

Allow me to place the concept of "inequity" and its relationship to emergency management in an historical context.

That remarkable species we call Homo sapiens (ourselves) has existed on earth for roughly 200,000 years, and we can trace the earlier ancestors to Homo sapiens back to about 5 million years ago. You may believe that or not, but the evidence is overwhelming. We also know that agriculture was first invented by humans on the order of 8000–12,000 years ago (depending on their geographic location). This means that fully-human beings have been able to grow their own food for only 5% of their existence; if you include the earliest human ancient ancestors, the figure is 2%. This in turn means that somewhere between 95% and 98% of human-ancestry had to earn its living by gathering, scavenging, and hunting, a way of life that requires small groups of somewhere between 35 and 50 individuals. After all, the natural environment provides only so much food that can be found in a day. Today, we have many cities populated by hundreds of thousands and millions, all because of agriculture.

The point is simply that, for the very vast amount of time we and our ancestors have been on earth, we utterly depended upon cooperative behavior within very small groups. Imagine, if you can, what it would be like to spend your entire life in a group of probably less than 50 individuals. I don't think we can really imagine that. We live among what is, in fact, a tremendous majority of strangers.

It is both necessary and easy to be empathetic and sympathetic to others in a small, highly interdependent group. It is so much easier to be self-centered among a large group of strangers.

Thus the long history of "inequity" is a telling characteristic of human societies. Thus do we enjoy the convenience of a highway and ignore the contaminants breathed in by local abutters. Thus do we ignore the danger of building residences in the wrong place and focus, instead, on the personal and municipal economics of rental properties. Thus do we not seriously concern ourselves (except, perhaps with short prayers and pittances of charity) with those decimated in so many ways by the loss of homes, jobs, and loved-ones. They are, after all, utter strangers.

Some inequities, of course, are simply due to the personal biases and prejudices of individuals and groups thereof; others are due to the long-term institutionalization of those same biases and prejudices. Fair to call the latter "structural inequities", that is, they exist because of the thoughtless carry-over from the past; the former, "personal biases" which exist because humans are very good at blaming others for their own misery.

Consider, for example, the number of women in the U.S. Congress over time: in any 10 year period from 1917 through 1967 from 0 to 17; and from 1967 through 2017, from 13 to 105. Still only 142 in 2021. Is this gender disparity due to women not wanting to become members of Congress? Highly doubtful, at best. And just why are women paid 70 cents per hour for the same work than men are paid 1 dollar? Is this gender disparity due to women not wanting equal pay? Even more doubtful, at best. No, it's because of a long-built-in structural aspect of a society in which women have been devalued; it's because society accepts without thinking what ancient history has passed on to them.

It is difficult to change either personal biases or structural inequities; both take much time to alter. Perhaps the main problem with personal biases is that they keep popping-up anew in each generation; with structural inequities, that they are most often camouflaged within the ever-expanding complexity of social systems.

I will have much more to say about the relationship between "inequities" and the future of emergency planning and response in Parts II and III of this volume. For now, suffice it to say that the rescue of individuals from roof-tops in a flood is essentially equivalent to caring for individuals who suffer the long-term effects of long-past emergency events. The two differ, quite simply, primarily in the manner in which we use the word "emergency" and collectively celebrate (or reserve) the appellation, "hero".

Chapter 2
Measuring Success and Failure of Response

How we measure success or failure in any endeavor we undertake essentially depends on the goals we have specified. In any emergency, there are many and diverse goals we may wish to achieve; we achieve some; others we do not. Both successes and failures result from a multitude of diverse factors, such as the adequacy of our planning and/or our execution, pure happenstance, and uncontrollable if not unforeseen circumstances and the torturous pathways of causes and effects.

In our personal lives, we often manage to deal with our individual failures by changing our goals. For example, I may want to become a physicist, but I keep flunking calculus exams...... so, I decide to become an artist. Fine, that solves that problem. Of course, such a solution is not appropriate for those professionals we depend upon for emergency planning and response, professions that require the recognition and subsequent correction of failures. In these professions, the first occurrence of a specific failure (due, perhaps to a blizzard that interrupts the delivery of necessary rescue or repair equipment or personnel to the site of an emergency) is understandable; a second occurrence of that same failure starts to become an unforgiveable failure in emergency planning. Failures, of course, are essential learning-tools!

I have been very fortunate in having benefited from the studies conducted by various governmental agencies (at both Federal and State levels) involved with different types of emergencies that analyze both the failures and successes of their response to emergencies within their jurisdictions. They do not tolerate the recurrence of past failures.

Figure 2.1 identifies some of the typical failures of response that are well documented. I first published this figure (Erickson, 2006) in a volume on Emergency Response Planning that relied heavily on information provided by the U.S. Fire Administration .... Information that is now about 17 years old. my intent is to demonstrate how they relate to issues raised in Chap. 1.

For example, all of these "failures" relate to specific emergencies (wildfires) and **specific operational aspects of actual response**. Today, these are standard

P. A. Erickson, *Effective Environmental Emergency Responses*,
https://doi.org/10.1007/978-3-031-05893-6_2

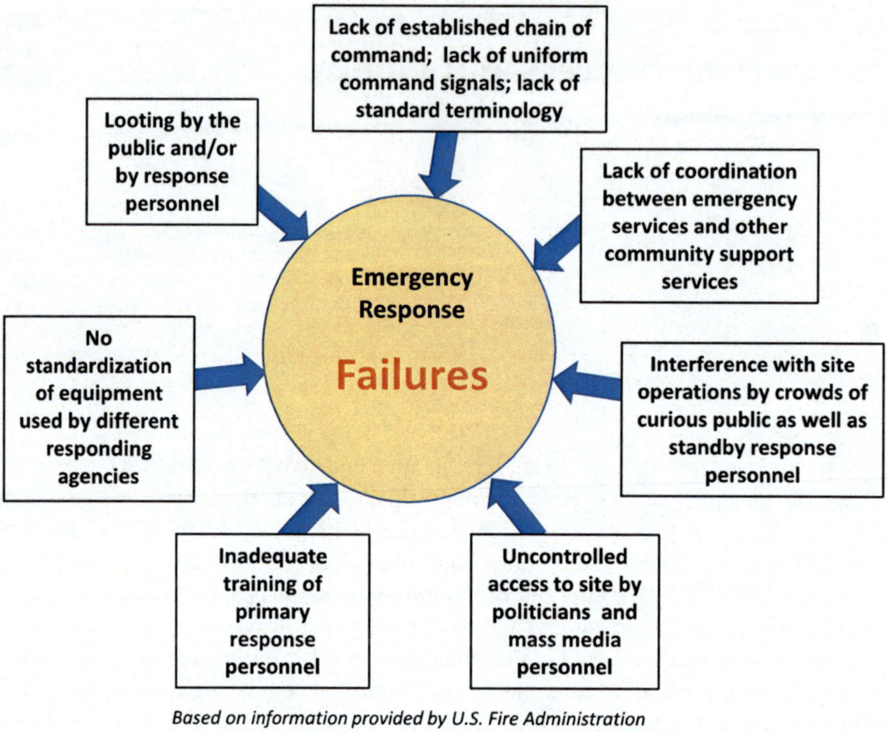

Fig. 2.1 Some examples of response failures. (Based on information provided by U.S. Fire Administration)

considerations in any planning and response regarding all emergencies. Note, they do not include any actions that directly relate to *preventive measures taken with regard to precursors* to the emergency, or *mitigations measures to be taken regarding long-term consequences*.

As noted in Chap. 1, there were no preventive actions taken regarding the denuding of unstable slopes in Malaysia or California.

And, as will be discussed in some detail in Chap. 11 (Terrorism), it has taken over ten tears to learn that on the order of 50% of first responders to the 9/11 terrorist attack have suffered debilitative effects or morbidity due to toxic inhalations of toxic debris particles. And what has been the follow-up of New York citizens who had to inhale the same debris? I hear a lot of silence.

While it is always necessary to consider past failures encountered in previous emergencies in the planning for future emergencies, it is also time to expand the scope of the very concept of "emergency response " to include appropriate considerations of failures of response due to the lack of meaningful attention to pre- as well as both short and long-term post-incident precursors and consequences. Such an approach is intrinsic to a more timely appreciation of the concept of "vulnerability", a qualitative measure of *susceptibility to harm due to exposure to hazard.*

In Fig. 2.2, I identify vulnerability in the context of the traditional equation, **Hazard** (*the potential for harm or injury*) + **Exposure** (*the amount and duration of exposure to the hazard*) = **Risk** (*the probability of experiencing the harm or injury*).

Please note that I have used a wide range of types of hazards, including "political" which I use to include decision-making in both partisan and other social-decision-making processes (e.g. private sector as well as governmental). I have also included broad ranges of time (under Exposure), and a very broad use of "Social Malaise" to include such issues as "economic distress", "looting and other crimes", etc. Hazard, exposure and risk are essential elements of any assessment of vulnerability which, as shown in Fig. 2.2, ultimately requires pre-incident evaluation of precursors and post-incident response as well as immediate response, both actions involving decision-making. I emphasize ……. making decisions (by whomever) play as much a role in the evolution of an emergency as it does in its response.

I'm reminded here of the expression, "What's good for the goose is good for the gander", a very old and laconic expression of equity between genders. I would very much like to add, "what's good for the first-responder is good for the first- and generational victim. For example, Figs. 2.3 and 2.4 include symptoms of **critical**

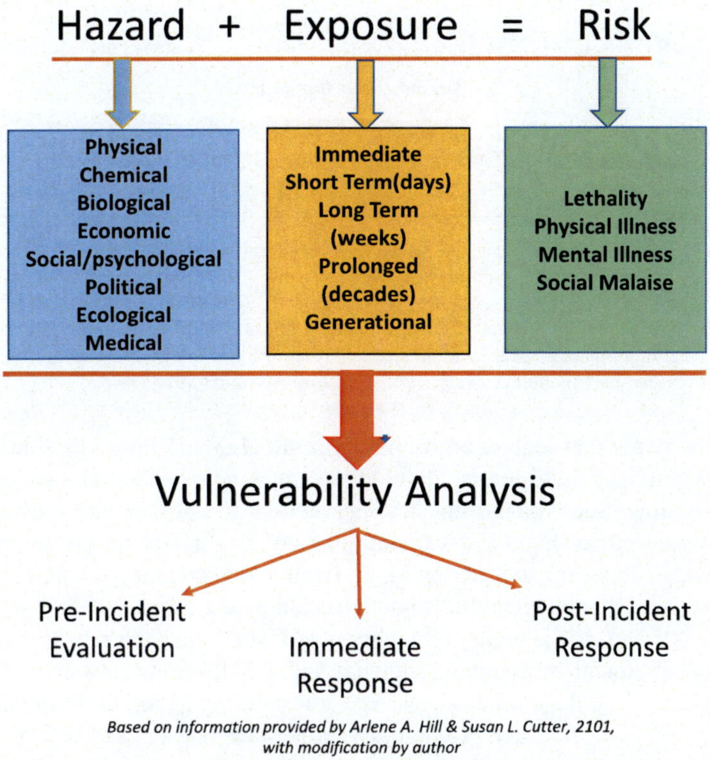

Fig. 2.2 Dimensions of vulnerability. (Based on information provided by Hill & Cutter, 2001, with modification by author)

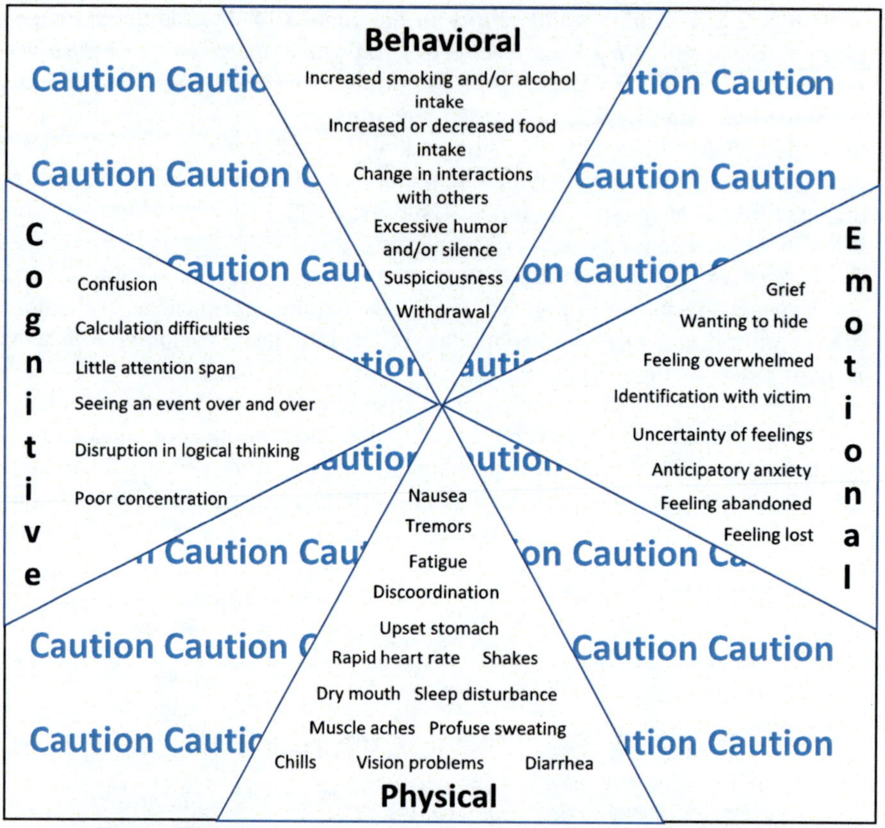

Behavioral, physical, cognitive, and emotional symptoms of critical incident stress that require cautionary monitoring (based on information provided by J. Mitchell and G. Bray, 1990, and U.S. Fire Administration, 1991

Fig. 2.3 Cautionary symptoms of critical incident stress. (Behavioral, physical, cognitive, and emotional symptoms of critical incident stress that require cautionary monitoring (based on information provided by Mitchell & Bray, 1990; U.S. Fire Administration, 1991))

incident stress that require **cautionary monitoring** and **immediate intervention** for first-response fire-fighters. Should they not also be applied to first- and generational victims? Such monitoring and intervention typically require substantial timeframes as well as high-cost expenditures. Who shall pay, and for how long? Somebody? To steal a quote from Mark Twain, "*I reckon they would, I dunno.*"

However, I do know that health-care (including the psychological care identified in Figs. 2.3 and 2.4) is highly variable in the United States in terms of availability, cost, and treatment-outcome. As shown in Fig. 2.5, there are substantial health-care inequities among the various states. Fair question .... again, as pertaining to decisions... how does the U.S. Congress and/or State Legislatures and/or Municipal Authorities square these inequities with certain entitlements in the Declaration of Independence? And once again, "*I reckon they would, I dunno.*"

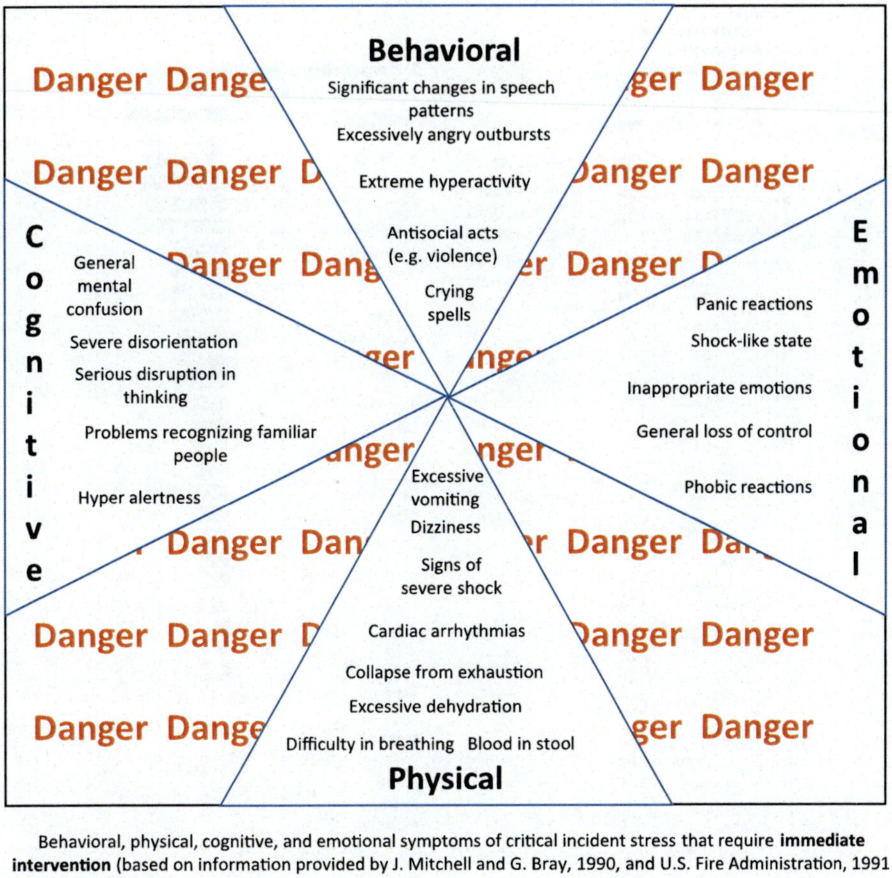

Fig. 2.4 Dangerous symptoms of critical incident stress. (Behavioral, physical, cognitive, and emotional symptoms of critical incident stress that require **immediate intervention** (based on information provided by Mitchell & Bray, 1990; U.S. Fire Administration, 1991))

But if they don't, I think it time they did.

At this point, I want to underline my persistent emphasis on the constant theme in this volume, and I beg some forgiveness for committing a perhaps too common sin practiced by authors…… namely, to quote myself:

> There is a parable told among the Eskimos of a hunter who wrestled a god. The hunter's wife had not conceived though she had performed all prescribed ablutions and, for lack of any hope that they would finally have a son, the hunter and his wife decided there was little reason to continue performing pious acts. One evening, Moonman sent is sled dog to howl outside their igloo. But so impious had the hunter become that he killed Moonman's dog. And when Moonman himself appeared and called out, "Who is it that has killed m dog?", the embittered hunter became even more insolent. He told Moonman to go away. When the god would not go away, the hunter wrestled him to the snow.

Table 2
2019 senior health state rankings and scores* listed alphabetically

2019 Rank	State	Score*
44	Alabama	-0.594
33	Alaska	-0.082
31	Arizona	-0.077
45	Arkansas	-0.652
17	California	0.256
5	Colorado	0.631
3	Connecticut	0.661
15	Delaware	0.284
29	Florida	-0.020
41	Georgia	-0.416
1	Hawaii	0.778
24	Idaho	0.115
31	Illinois	-0.077
36	Indiana	-0.240
13	Iowa	0.401
28	Kansas	-0.019
49	Kentucky	-0.880
48	Louisiana	-0.857
8	Maine	0.448
12	Maryland	0.406
11	Massachusetts	0.411
26	Michigan	0.058
4	Minnesota	0.655
50	Mississippi	-0.989
39	Missouri	-0.300
27	Montana	-0.001
23	Nebraska	0.146
38	Nevada	-0.298
6	New Hampshire	0.552
21	New Jersey	0.178
42	New Mexico	-0.454
16	New York	0.280
37	North Carolina	-0.272
18	North Dakota	0.233
30	Ohio	-0.063
47	Oklahoma	-0.784
19	Oregon	0.194
19	Pennsylvania	0.194
7	Rhode Island	0.456
34	South Carolina	-0.102
22	South Dakota	0.162
43	Tennessee	-0.528
40	Texas	-0.318
2	Utah	0.710
8	Vermont	0.448
25	Virginia	0.075
14	Washington	0.396
46	West Virginia	-0.687
10	Wisconsin	0.444
35	Wyoming	-0.166

*Weighted standard deviation relative to U.S. value. A score of 0 is equal to the U.S. value.

Table 2 lists the 2019 senior health rankings alphabetically for all 50 states. In Figure 7 green bars represent states scoring above the U.S., while yellow bars represent states scoring below the U.S. Washington (No. 14) and Delaware (No. 15) have the largest difference in score, making it difficult for Delaware to move up in the rankings. There is also a large gap in score between Mississippi (No. 50) and the next state, Kentucky (No. 49).

Figure 7
2019 senior health state scores* listed by ranking

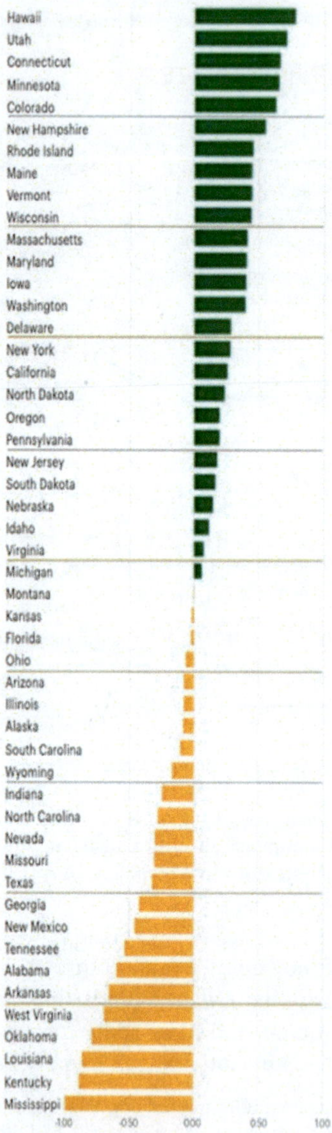

* Weighted standard deviation relative to U.S. value. A score of 0 is equal to the U.S. value.

AMERICA'S HEALTH RANKINGS® SENIOR REPORT
United Health Foundation, Senior Report 2019

Fig. 2.5 Disparaties in American health care

> The battle lasted many hours. However, the hunter fought like a wild animal and was finally able to get the better of Moonman, who, after all, thought he was fighting a mere man. In the end, the hunter was about to kill Moonman when Moonman shouted in desperation, "What? Then there will be no more ebb and flow of the Sea! And there will be no more baby seal!" At that, the hunter became human again. He let Moon man up and promised to do his rightful penitence.
>
> To any student of anthropology this story is now new. It is told in different ways, with different names, in different cultures throughout the world. Yet it is always the same. In societies where people live with a thin insulation between themselves and their physical environment (and that includes all societies) , the moral is clear – the measure of humankind is neither power nor the lack of it, neither plenitude nor the lack of it, and certainly neither peace nor lack of it. But the measure of humankind is the acceptance of the ultimate and unalterable dependence upon environmental processes and cycles. **Our contemporary rediscovery of this truth is causing us to reconsider how we perceive our environment and what we do with it. In the process, we are changing the ways by which *complex bureaucracies make and implement decisions*.** (Erickson, 1979)

What does this have to do with emergency planning and response. Everything!

Decisions (those made and those not made) are as often the precursors to emergencies as are plate tectonics, hurricanes, wildfires and any other environmental phenomenon or process. Simply put, decisions have consequences!

Who made the decision to denude the hillside in Malaysia (mentioned above) and/or build a condo at the foot of that hill? Who considered that the upland slope along Big Sur (also mentioned above) was or was not likely unstable due to a previous wildfire? Who delayed an evacuation notice prior to the landfall of Hurricane Katrina (Chap. 9)? Who decided it was time to "return to normal" in various states in the midst of the Covid 19 pandemic (Chap. 5)?

It is not that we have never paid attention to the likely consequences of making decisions regarding our environment. In fact, Congress passed the National Environmental Policy Act in 1969 (NEPA), and over a dozen states enacted comparable legislation.

While NEPA pertains directly to Federal projects (e.g., interstate highways, pipelines, etc.), State environmental laws show a very interesting diversity. Some require an environmental review by local government projects (California, Georgia, Minnesota, New York, Washington); some (California Minnesota, New York), by private individuals and businesses (e.g., agricultural, or other projects that require state permits or funding); some (California and New York), by state authorities having jurisdiction over projects affecting climate change.

The purpose of these Federal and State laws is *to assure that governmental decision-makers give proper consideration to the environment prior to undertaking any major action that significantly affects the environment.*

As shown in Fig. 2.6**,** NEPA is a process involving the assessment of proposed actions on both the physical and social environments; this assessment includes an evaluation of both positive and negative impacts and requires a subsequent mitigation of negative and an enhancement of positive impacts. Yes, a typical NEPA is most often expensive and time-consuming. I am not recommending that proper emergency planning and response include a formal adoption of NEPA. *But I do*

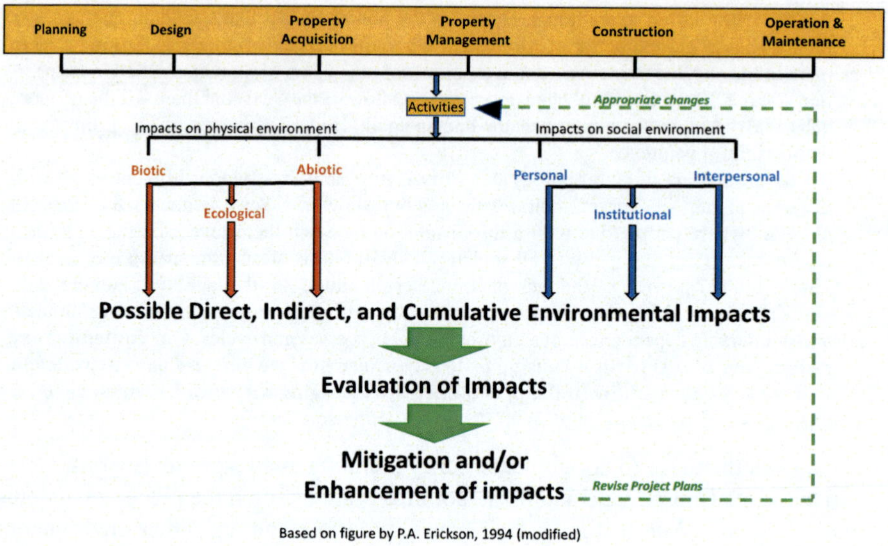

Fig. 2.6 Overview of proposed project under national environmental impact assessment process. (Based on figure by Erickson, 1994 (modified))

suggest that the ethos of NEPA should be considered in achieving a comprehensive, effective emergency management process that includes consideration of (1) precursors to emergencies and (2) both immediate and long-term, equitable remediations. Such an approach would obviously require serious revamping of governmental and private-sector interactions as well as significant societal consensus (see Part III of this volume).

A word or two about "revamping decision making"……. not an easy thing to do. Institutions (whether governmental, private sector, charitable, etc.) always have a history behind their rules and protocols…… their way of doing things. Changes are often disruptive to some degree and sometimes outright anathema to those used to doing things in a particular way. For example, State Police and Municipal Police in the same State sometimes have somewhat different protocols for directing/diverting traffic in the midst of an emergency; both are well experienced and perform their duties admirably; but sometimes they just plain "disconnect".

It may also happen that different rules (more specifically, governmental red-tape) exacerbate problems between emergency-related personnel procuring debris-removal and other services and local private-contractors. Such problems can rapidly escalate, with private contractors becoming increasingly obdurate regarding their cooperation with emergency-related-authority.

And so, on and on. Effective emergency planning absolutely requires a smooth integration of all appropriate protocols and rules. If that requires a revamping of historically contrived and individualized institutional procedures, so be it! After all, it's the twenty-first Century, not the twentieth.

The ethos, the intent …. the goal of *Impact Assessment* of both the physical and social environments in order to prevent and/or mitigate negative impacts of human decisions …. isn't that goal and intent consistent with emergency planning? Obviously, I think so.

References

Erickson, P. A. (1979). *Environmental impact assessment: Principles and applications*. Academic.

Erickson, P. A. (1994). *A practical guide to environmental impact assessment*. Academic.

Erickson, P. A. (2006). *Emergency response planning for corporate and municipal managers* (2nd ed.). Elsevier.

Chapter 3
Logistics: Expertise, Personnel, Equipment, Supplies

In a conversation with a friend of mine who had a long experience in the military, he told me, "Well, Paul, …. Logistics comes down to this …. getting there the fastest with the mostest!". Allow me to translate this typically laconic army-speak: the word "there" means *wherever* there is a problem that needs fixing; the word "fastest" means "well *before* the problem becomes evident; the word "mostest" means "with *more-than-enough of whatever is or may be necessary*" to deal with the problem. At the time, I thought this view of logistics was supremely comprehensive. But that was more than half-way through the twentieth century; we are now well into the twenty-first.

Back then, we were still basking in the allied success in World War II and, in many ways, it was a time in which we were dazzled by the ever- increasing mechanization and novelties of our industrial productivity that had almost-explosively evolved from the early 1800s onward to a nation built on massive and increasingly pervasive consumption.

Today, the issue is not simply national productivity but, rather (and essentially), both the National and global distribution of what is produced.

As of 2020, there were roughly 3000 satellites in orbit, with almost 50% if them being devoted to communications used by governments, corporations, private-sector organizations, and individual persons, all of whom have 24 h. access to digitized information and communications transmitted at the speed of light throughout the globe ……information and communications that interconnect with every social-sector having emergency-related capabilities regarding expertise, personnel, equipment, and supplies.

However, with regard to the ease-of-access to and the employment of logistic-necessities, one does have to take into consideration a basic fact that underlies what once was and yet remains a significant characteristic of the United States. While we wax eloquently on the phrase "E Pluribus Unum", or the vow of The Three Musketeers ("All for One and One for All!"), we cannot ignore a fact which James

P. A. Erickson, *Effective Environmental Emergency Responses*, https://doi.org/10.1007/978-3-031-05893-6_3

Madison, the major Founder of our American Constitution so elegantly and forcefully examined…. namely, the proliferation of factions.

> The latent causes of faction are thus sown in the nature of man; and we see them everywhere brought into different degrees of activity, according to the different circumstances of civil society. A zeal for different opinions concerning religion, concerning government, and many other points, as well of speculation as of practice; an attachment to different leaders ambitiously contending for pre-eminence and power; or to persons of other descriptions whose fortunes have been interesting to human passion, have, in turn, divided mankind into parties, inflamed them with mutual animosity, and rendered them much more disposed to vex and oppress each other than to co-operate for their common good. So strong is their propensity of mankind to fall into mutual animosities, that where no substantial occasion presents itself, the most frivolous and fanciful distinction have been sufficient to kindle their unfriendly passions and excite their most violent conflicts. (Madison, 1887)

While there can be no question as to Madison's political brilliance or his mastery of logic, he was nonetheless (as are all of us) "a child of his times". Having presented his prescient analysis of the causes and evils of "factions", and proposing their preventives, he focused on two aspects of the proposed new Republic we call America, namely, its growth in population, and its already even-then -evident growth in physical extent. I quote him once again from Federalist No.10:

> Extend the sphere, and you take in a greater variety of parties and interests; you make it less probable that a majority of the whole will have a common motive to invade the rights of other citizens; or if such a common motive exists, it will be more difficult for all who feel it to discover their own strength, and to act in unison with each other.

Of course, as a child- of-his-time, there was no way that he could even imagine a future when horses and boats with their weeks- to-months-long travel-times would be replaced by travel-minutes and hours and instantaneous communications between even most distant like-believers in whatever cause. Within a mere 80 years in his future, but even then with new travel- and communication-technologies, The Civil War gave a profound rebuttal to his assurance regarding the control of factions. And the United States has experienced many instances of severe rebuttals of Madison's thesis from that time up through the present, with the result that the very word "faction" has commonly come to denote a negative aspect of social interactions.

Yet, the word "faction" simply denotes a smaller group within a larger population and, as Americans, we do celebrate diversity which, after all, includes the concept of dissention (or disagreement in interests, values, as well as diverse behavior). Yes, we like to think there is such a thing as "American culture" or "American society". Culture refers to learned behavior, but does that mean all Americans learn and believe precisely the same things and daily behave the same way? Likewise, the word "society" refers to a human group within some geographic boundary, but does that mean all Americans are the same, that they have the same religion, the same wealth, the same education, the same experience, or the same-anything?

Of course not. The words "culture" and "society" are outworn nouns contrived much more of romantic wishing than of demonstrable fact. However, the adjectives "cultural" and "social" are much more meaningful because they describe what is learned and what is learned by virtue of the interactions of groups and individuals.

Decisions made in the progress of emergency planning and response obviously must make full use of modern communications among the various societal resources of expertise, personnel, equipment and supplies, but as shown in Fig. 3.1, these resources must be seamlessly coordinated in a *persistent atmosphere of social and political factions that, despite their many differences, depend directly and/or indirectly on the vagaries of both collective and personal economics.* Simply, money counts!

Of course, we have a long- enough experience to understand that any phase of logistics-in-motion is fraught with a wide variety of problems. Governmental agencies as well as private businesses have given and continue to give careful attention to just what those problems are and how they can be resolved. But money still counts!

Do we spend it on managing the precursors of emergencies or on repairing the damage they cause? If the latter, do we spend on the short-term physical damages alone, or do we also spend on persistent, long-term deprivations of the health and security of victims. Where is the consensus among current factions? No, political and social consensus is always very hard work...... and, sometimes quite perilous.

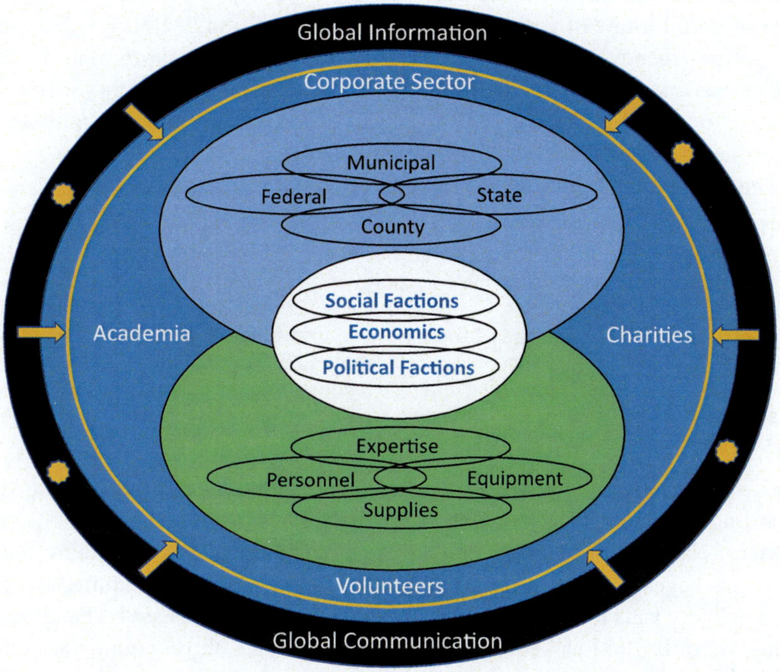

Fig. 3.1 Global elements of supply-chain-logistical success

In its "Federal Response to Hurricane Katrina: Lessons Learned" (U.S. Government Printing Office, 2006a, b), the Task Force convened per order of the President noted as a **Critical Challenge**, the following:

> The scope of Hurricane Katrina's devastation, the effects on critical infrastructure in the region, and the debilitation of State and local response capabilities combined to produce a massive requirement for Federal resources. The existing planning and operational structure for delivering critical resources and humanitarian aid clearly proved to be inadequate to the task. *The highly bureaucratic supply processes of the Federal government were not sufficiently flexible and efficient, and failed to leverage the private sector and 21*st *Century advances in supply chain management.*

This willingness to admit failure is laudatory. After all, we learn most often from our failures. But that the above-noted "inadequacy" is attributed to (1) failure to leverage the *private sector*, and (2) failure to utilize twenty-first century advances in *'supply chain management'*, is even more telling, especially with respect to the adequacy of the integration of modern technologies and procedures with the realities of emergency-response.

Old ways to doing things are sometimes (often?) counter-productive at best. Yet (regrettably) old ways of doing things tend to become institutionalized… formalized and fossilized as inviolate protocols.

For example, there are roughly 30,000 fire departments in the United States; about a fifth of them are fully or mostly career staffed. Now, how many provide their personnel with regular (hopefully, yearly) medical checkups for lung damage due to smoke inhalation and for other diseases (e.g., cancer, heart disease, etc.)? I don't have (nor do I know of anyone who has) the precise number. NFPA (National Fire Prevention Agency) 1582 requirements are designated as appropriate for all firefighter classifications – career, volunteer, and industrial firefighters (full-time, part-time, or on-call). However, NFPA has no enforcement authority, and some fire departments nationwide do not follow NFPA 1582.

The following quote is, at the least, unnerving:

> The economic costs associated with a firefighter's death far outstrip the costs of ensuring that firefighters receive an annual medical examination. Put simply, firefighters' lives are worth far more than the cost of implementing a wellness-fitness program that starts with an annual physical. Knowing this, it is the position of the IAFC that every firefighter receives an annual medical examination for the purposes stated above, and that this examination should follow as closely as possible the guidance of NFPA 1582. Fire *Chiefs have an obligation to find funding sources and develop creative strategies* to ensure the safety of their personnel. This document is designed to support that responsibility *(International Association of Fire Chiefs).*

Not to leave out the private-sector; another example from the current situation regarding Covid 19 Isolation: One consequence has been the proliferation of home services, especially the 1-day deliveries of packages of most anything, including heavy packages, such as furniture and other heavy items that require heavy lifting from delivery vans to the front door. Most of those deliveries are by hand, with little if any use of lift-and-roll equipment. I keep thinking if those young backs and hips and legs and arms after a few or years of manual lifting of over 50 lbs. when, in fact, there is a wide variety of off-loading devices that minimize muscle-tare and vertebral compression and are easily available. Oh, well…we've always done it that way!

Yes, but I have also noticed that many garbage-collection trucks have converted to trucks outfitted with modern automatic equipment for lifting and on-loading garbage cans. Drivers don't have to get out and do the heavy lifting. Good for them. Good for their companies.

I do not intend to argue that the newest is always the best, nor the oldest is always the worst. A well-informed opinion as to what is the preferable way to do something is best arrived at by comparing and assessing both the advantages and disadvantages of older and newer ways of doing things. I suggest the following mental exercise that I personally find useful.

Take a time period and sequence of years…. say, from 1925 to 2025 in 20-year increments…… thus, 1925, '50, '75, 2000, '25. Now do some homework (either with a history-book or the internet) and choose key characteristics of the U.S. (e.g., technology, population size, mean family income, level of education, medicine, etc.) for each of those 25-year periods (25 years being a good number between human generations). Then, if you want, decide in which generation you would want to live. Myself, I find 1925 to be intellectually exciting, but I wouldn't like to live back then. The level of knowledge and technology was pretty much negligible as compared to today.

Well, that's because I grew up in the '50s and '60s; a lot of changes since the 20s, but no thanks; modern technology (transportation, medicine, the sciences) was in its infancy. You get the point. Choosing which previous (as opposed to the present) generation you might prefer would mean living in the midst of much less knowledge, more limitations to travel, education, recreation, employment, and health than the present. Perform the same exercise 25 years from now and you will come to the same conclusion. It doesn't mean the past is bad……. simply, that the past is always more limited in possibilities, and the present is typically on at least the threshold of a much greater range. Nonetheless, there are valuable examples of ancient knowledge that would serve us well to keep very much in mind. You just have to be selective but being selective means you must know what that old knowledge was and how it was used. That depends on the thoroughness of your inquiry.

Old knowledge can have direct relevance to a recurrent contemporary type of emergency. For example, today, we suppress fires and consequently allow the accumulation of deadwood and dry brush which serve as fuel for fires hot enough to burn trees. Indigent Americans (better named 'First Peoples') burned the deadwood and brush twice a year, thereby removing the risk of fires hot enough to ignite thick barked mature trees. Colonists were amazed by the fact that they could actually gallop through forests denuded of underbrush (Taylor, 2001).

Well, of course, the First Peoples of North America had many thousands of years' experience with forest management before the arrival of those post-Columbus, European immigrants we mistakenly refer to as "first-Americans". They knew precisely how to manage fires. And, from the seventeenth century through the nineteenth, so did most Americans.

A personal example: my first-owned house was built in 1804; it had six fireplaces, because wood fires were the only source of heat. Back then, you had to know how to light fires without burning-down the house. But now …… a good guess is that probably 90+ percent of Americans haven't the slightest knowledge of how to maintain a safe wood-fire. Today, we're very good at pushing buttons to get our heat. And, today, we're the primary campground-incendiaries of our forests at the cost of billions of dollars and many lives.

Other examples of effective old-ways of doing things (in this case, more specifically, distributing resources from those who have them to those who do not) is given by native North Americans (northwest coast) and the!Kung San (western Kalahari Desert): the first is a ceremony (known as Potlatch) in which possessions are given away to others (or destroyed) to display wealth or enhance prestige; the second is also a ceremony (known as Hxaro) in which gift-giving and sharing establishes a mutual-reciprocity between social groups that minimizes subsistence risks during difficult times. Today, Social Anthropology is increasingly focusing on how even long- extinct human ancestors might also have handled the persistent and omnipresent need for different social groups to provide vital support to each other (Barnard, 2011).

The point is that, from our earliest ancestors throughout human history, there are social groups that have, at any one time, had to struggle through very difficult times, with the consequence that humans have contrived many ways to provide for each other with the necessary means for survival and, in so doing, have established mutual assistance as a hallmark of social consciousness. Yes, in our contemporary lives, it is oftentimes difficult for us to recognize the reward or pay-back for our altruistic behaviors on behalf of those so many millions who are essentially strangers. I suggest that our payback is, at the very least, a vibrant, productive sense of social-solidarity instead of the numbing sense of destructive social-anomie.

3.1 Expertise

There are many kinds of expertise (i.e., a merging of comprehensive and authoritative knowledge with extensive experience), and each can be found in numerous and different components of any social system. Expertise is not confined to a particular small group (as it was in ancient times), but proliferates today in diverse institutions, organizations, and individuals.

For example, in a particular emergency (e.g., flood), I may be concerned with the spread of disease by insects or aquatic parasites. Yes, there are Federal agencies that can give me useful information; so can many State biologists, and they, most likely, have a more relevant knowledge of local bio-populations (or maybe not). I once knew a long-practicing physician also was otherwise (by a long-term hobby) an expert ornithologist who was typically consulted by State environmentalists concerned about the effects of certain land-management practices on avian populations. Who "woulda thought?"

And what of academia, the corporate sector, charities, and even (having above mentioned a multi-talented physician) individuals? The issue is not any lack of expertise. The issue is identifying the expertise needed and then arranging that collective expertise into an efficient and effective management team. Such, of course, is the stuff in which experts in supply-chain-management in military and, in this generation, corporate organizations) excel.

One of the common problems in emergency planning and response directly related to any pool-of-expertise is the *practical application of that expertise*. For example, as will be seen in Section II, a few common emergencies (e.g., floods, wildfires) typically involve subsequent risks due to contamination of properties, soil and water supplies with disease-bearing microbes, parasites, and insects. Health related agencies (Federal and State) certainly have the expertise needed by victims' intent on cleaning-up their properties as quickly as possible. However, the typical citizen doesn't know what specific protective clothing is required, or how to obtain it, how to properly don that clothing or properly remove it, nor how to safely dispose of contaminated debris, decontaminate their shovels and brooms, or safely dispose of their protective clothing.

Expertise that is not clearly transformed into practical application is useless. One way to ensure practical application of expertise is to provide citizens with (a) emergency personnel adept in "show-me-how" demonstrations and, (b) the protective material and equipment needed; another is to provide commercial cleaners who have the appropriate tools and know-how. Are we willing to provide the cost? By which political or private (e.g., insurance) entity?

Another obvious (and always present) problem relates to the crucial embedment of expertise in the planning phase of Emergency Response.

I recopy the first sentence of a previous quote from the Federal Response to Hurricane Katrina: Lessons Learned: "The existing ***planning and operational structure*** **for delivering critical resources and humanitarian aid** ***clearly proved to be inadequate*** to the task.

My proclivity has always been to consider "personnel, equipment, and supplies" as being the "muscle tissues" of emergency response, while "expertise "was the heart and soul" of emergency planning. Without an adequate plan, there is no operational structure for implementing a successful response, regardless there being even a profusion of equipment and supplies.

3.2 Equipment and Supplies

Most recently, the Covid 19 epidemic provided two examples of logistic failures: one regarding the production of supplies and, one regarding their distribution.

Failures in the production of personal protective equipment and supplies needed by essential personnel in hospitals, ambulance-transport, laboratory personnel needing chemical reagents to complete testing for viral infection, and other community first-responders daily exposed to possible viral infection. By any measure, the shortage of such supplies was, to say the least, unbelievable, and utterly despicable in such a technologically advanced nation as the United States. The cause of this catastrophe was decision-making (specifically, the non-making of appropriate decisions). Even when such supplies did become increasingly available, their transport and distribution were not based on or managed according to well established protocols of supply-chain-management but, shall we say, "other considerations". Given the potential inadequacies of any decision-making process (including lack thereof), it would seem reasonable to concentrate on the construction and periodic resupply of stockpiles strategically located for close regional (and alternative) transport and direct relevance to likely vulnerabilities of local communities.

There have been other recent failures regarding the timing of arrival of shelters for citizens displaced from their homes due to geographically extensive hurricanes, tornados, and floods. The adequacy of temporary shelters, even when available, is often questionable, and very often are not sufficient for relatively long periods of time which, for low-income and/or otherwise marginalized populations, become basically hovels where people subsist as best they can on their own, without benefit of income, health-care, or even nutritious food and clean water. And, of course, they can quickly become targets of the whims of criminals and psychopaths.

Another form of shelter is, of course, a hospital or other medical facility. Again, the current pandemic has highlighted the fact that hospitals over-run by covid patients cannot service normal health-care services, such as optional surgeries, early diagnostic tests for various diseases (including cancer, where early intervention is so important).

The Army Corps of Engineers has demonstrated how to turn other buildings (unused depots, performance theaters, etc.) into functional hospitals in a matter of days …. again, a good example of the military's expertise at meeting the needs of immediate emergencies. And the corporate world has also learned from the pandemic that they need not rely of the physical presence of their employees (who, after all) have been confined to their homes because of Covid; in fact, many companies have found that a virtual workforce is typically more productive than an office-bound one. Thusly do we sometimes learn that emergencies can provide potentially useful ideas to consider for the future.

If I may be (as is my usually proclivity) simple-minded, I'd say, at this point, "that if it takes a village to raise a child", it takes all the components of a society to find the best way to deal with emergencies. A single institution (whether governmental, corporate, or whatever other term you may use to dissect a social system) cannot do it alone.

3.3 Persons: Emergency Victims

We have the muscle-power; we have the collective-expertise; we have the more than sufficient productive-capacity for whatever equipment and supplies may be required. The measure of successful response, however, is very often determined by the decisions made (or not made) in the use of resources to prevent and/or ameliorate the harm to those persons that any emergency entails, *including not only those who are immediately endangered by an unfolding emergency (henceforth, I will refer to them as "first-victims", but also those whom we call "first-responders". Both are, in fact, victims* (Fig. 3.2).

Even as jointly characterized as victims, there are, nonetheless some basic differences as well as similarities between both.

For example, I recently listened to a series broadcast interviews which included a family who had lost their home to a flood. Both husband and wife tearfully explained how that home, built on a floodplain had been in their family for four generations and loudly announced their intension to rebuild. Yes, and be flooded out again……and, of course, have to be rescued again by first-responders who have to risk or own lives again …. and again.

An obvious question arises: do Americans have a constitutional right as homeowners to place other Americans at risk-of-life by persistently building on flood plains?

Who should make that decision? That question is not any different from the question raised by the Covid-19 Pandemic: do Americans have the right to infect other Americans by not wearing facemasks or keeping social-distance? Nor does either question square with the historic American understanding that yelling "Fire!" in a crowded \theater (which is not then on fire) or not yelling "fire!" when there is one cannot be tolerated.

So, in this case, people who choose to live on floodplains (regardless of there being levees, which often fail) are, in effect, choosing to risk floods. And they can do so because a variety of decision-makers (e.g., politicians, mortgage bankers, insurance companies) allow them to do it. Such decisions by all parties concerned are clearly acting as precursors to subsequent flood-emergencies.

On the other hand, first-responders also choose their professions and willingly take the risks attendant thereto. But they do that daily, at all hours … continually

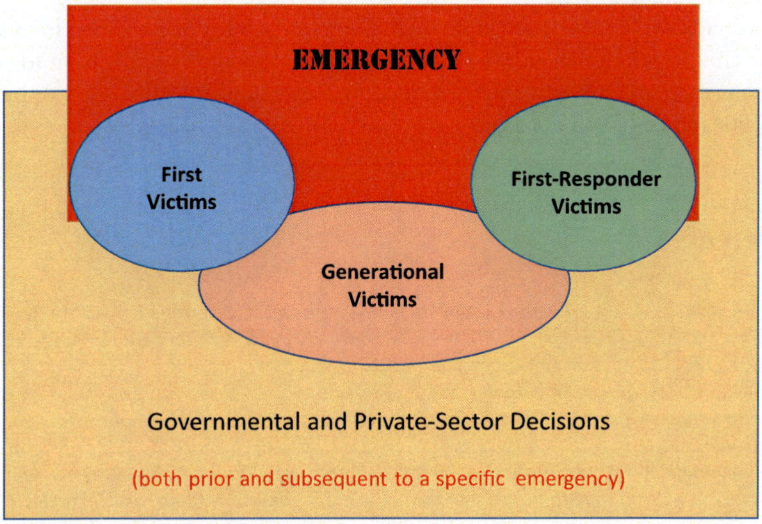

Fig. 3.2 Decisions as precursors and palliatives

over years exposing themselves to physical and mental risk. For the flooded-out first-victims, physical and mental risks are typically (thankfully) episodic and, but also physically and mentally devastating.

Some physical injuries require substantial medical care, and psychological injuries typically require long-term care of both first responders and first-victims by psychological councilors. How do they pay the costs? How long do they qualify for aid and assistance? What succor does our social system offer, and just what changes might be seriously considered as correctives to persistent inequities?

As shown in Fig. 3.2, emergency victims (first victims and responders) obviously experience traumas that most of us can only imagine; whether it be the daily risk of responders or the utter horror of suddenly losing home and/or loved ones, the toll is devastating and has lasting effects which, in turn, can lead to familial disfunction, substance abuse, sleep-deprivation, prostration, even suicide, each of which, in their turn, may have lasting psychological effects on other family members, especially children. I refer to such sequential effects as "generational" … resulting in not only long-term psychological but also physical debilitation (e.g., as in radiation poisoning from nuclear meltdown emergencies). Where does one go for help if one has lost a livelihood, exhausted savings, been denied credit, and been evicted? Or, is there a time-limit to emergency response? Perhaps a time determined by the newsworthiness of the emergency, …. or by the practical attention-span of politicians, …. or the diversion of public funds to a more publicly popular bauble.

One last example: I recently read a news article published by the BBC regarding a family in Florida (Padhy, 2020). A father of three children lost his job during the covid 19 pandemic, leaving him to live off his savings in a motel. Four weeks behind in their rent, they were faced with eviction. The motel's eviction policy allows items deemed non-essential can be removed from the room, which included the TV. The youngest son (6 years old) is autistic; unable to speak, the TV is the primary means of keeping him calm.

So much for "E Pluribus Unum". So much for "All for one and one for all". Well, these are old principles, but they are still 'golden goodies'. We'll have to see what happens to them in this twenty-first century. Right now, it's all so easy to ignore the inequities imposed on strangers.

References

Barnard, A. (2011). *Social anthropology and human origins*. Cambridge University Press.

Madison, J. (1887). Federalist No. 10. In *The Federalist Papers*. Country House Publishing (2015).

Padhy, K. (2020, December). U.S. eviction crisis. *BBC*.

Taylor, A. (2001). *American colonies*. Viking Press.

U.S. Government Printing Office. (2006a). *Federal response to Hurricane Katrina: Lessons learned*.

U.S. Government Printing Office. (2006b). *A failure of initiative: Final report of the select Bipartisan Committee to investigate the preparation for and response to Hurricane Katrina*.

Part II
Lessons Learned; Lessons Applied

1.1 Preface to Part II

The following chapters focus on familiar types of emergencies. The intent of these chapters, however, is not to recite all the voluminous historical details of specific emergencies. Such information and data are easily available from both printed and digitized informational sources. Of course, there are, nonetheless some selected issues that deserve attention. The primary intent is to select certain aspects of each that underlie and expand upon those concepts and considerations presented in Part I.

Each chapter contains a summary diagram that depicts various interconnected causal-chains based on recent experience and (sometimes) theoretical possibilities. However, I do not claim that these diagrams are by any measure fully-inclusive.... they should be viewed as heuristically useful examples.

The text of the following chapters gives specific attention not only to interconnected causal-chains (both in precursor and subsequent consequences) but also to decision-making processes that are crucial to both emergency planning and response, including political and private-sector decision-making in the context of American social dynamics, especially with regard to institutionalized- as well as persistent, factionally-invigorated social-inequities. In this regard, my primary focus is on the United States; of course, in this global village of ours, lessons learned in one nation are quite often pertinent (to some greater or lesser degree) to others.

Provided at the end of each chapter are some exercises that I trust will be helpful to the reader for comparing key similarities and differences in the causes and effects of the various types of addressed emergencies. Teachers may find these exercises useful for class examinations or for written assignments (whether graded or not) and/or for oral presentations.

Finally, At the end of Part II, I am including a bulleted summary of points I want to emphasize as preludes to the discussions presented in Part III.

Chapter 4
Wildfire

Given the characteristics of a wildfire, perhaps the word "hellfire" is more appropriate. With flame heights of well over 150 feet and temperatures of over 2000 °F not to mention wind gusts up to 100 miles per hour, and the explosive incineration of both human and non-human habitations, wildfires qualify as the near-perfect representation of Hell on earth. However, not to obfuscate the distinction between divine and secular causation, I will hereinafter retain the term "wildfire"; after all, we know for a fact that about 90% of wildfires are due to human behavior (e.g., campfires, careless disposal of cigarette butts, cook-stoves, etc.), while only about 10% are due to lightening.

As shown in Fig. 4.1, the annual number of acres burned in the U.S. (millions) has generally increased from 1991 to 2020, while the number of fires (thousands) has decreased. How interesting! That apparent trend obviously raises some questions.

Are forests and other biomes becoming intrinsically more vulnerable? Does it have anything to do with human access to more vulnerable ecosystems? Maybe an improved technology regarding cook-stoves and portable fire-cauldrons? Is it some synergistic interaction of these and other possible factors? Or, Is the apparent trend meaningful at all?

Extending the data back to 1980 by using additional information (Fig. 4.2), we see more specific variations in annual numbers of burned acres. These data show that the overall average of burned acres from 1980 through 1999 is a little over 3 million; from 2000 through 2020, a little over 7 million, about a twofold increase. This pattern is reinforced by Federal cost-figures for wildfire suppression (costs to contain or suppress fires) that rose from $ 425 million in the period 1985–1999 to $ 1.6 billion in 2000–2019 (Roman et al., 2020). Even the incidence of residential-building (i.e., homes and apartments) cooking-fires increased by 10% in the period 2010–2019 (U.S. Fire Administration, 2021). While not necessarily relevant to wildfires, this statistic would appear to underscore the evident proclivity of modern humans to mishandle combustibles whether they be in an apartment or in a forest.

P. A. Erickson, *Effective Environmental Emergency Responses*,
https://doi.org/10.1007/978-3-031-05893-6_4

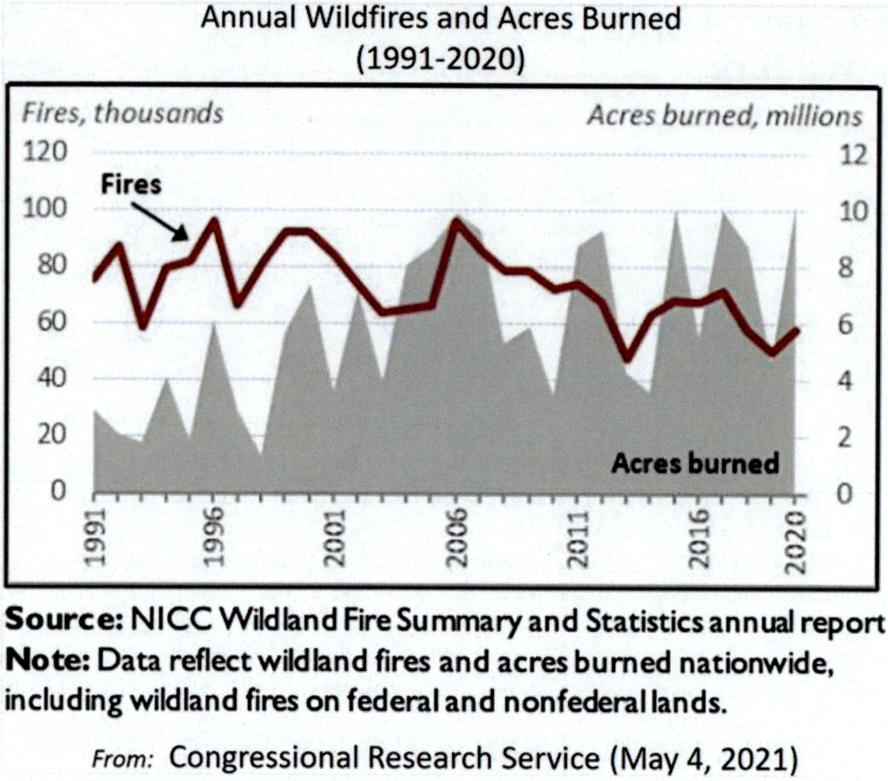

Fig. 4.1 Annual wildfires and acres burned (1991–2020). (From: Congressional Research Service (May 4, 2021))

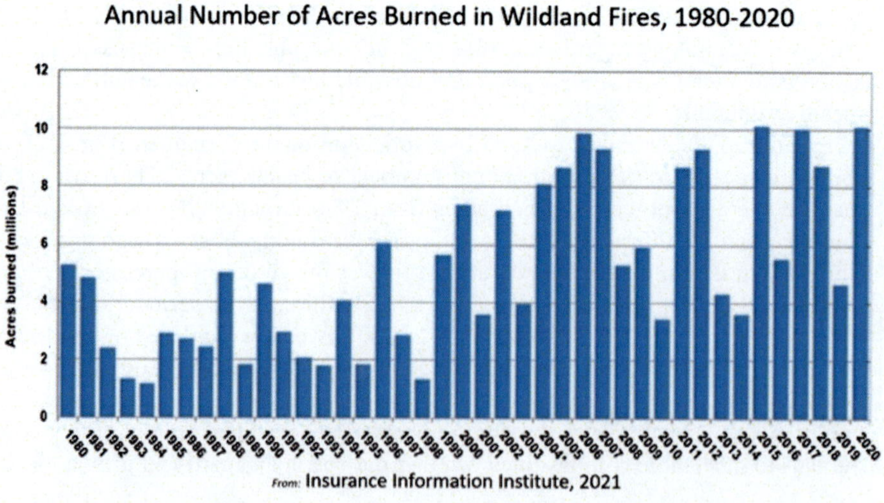

Fig. 4.2 Annual number of acres burned in wildland fires, 1980–2020. (From: Insurance Information Institute, 2021)

Given differences in climate, soils, and vegetative cover, different States are, of course, more or less vulnerable to wildfires (Table 4.1). However, wherever they occur, wildfires typically incur substantial monetary costs, not only to the Federal Government (depending on legal requirements between Federal and State agencies) but also to State and municipal agencies.

Table 4.1 Number of wildfires by state, 2020

Alabama	836
Alaska	349
Arizona	2524
Arkansas	55
California	10,431
Colorado	1080
Connecticut	586
Delaware	426
Florida	2381
Georgia	1699
Hawaii	58
Idaho	944
Illinois	19
Indiana	11
Iowa	126
Kansas	52
Kentucky	524
Louisiana	401
Maine	1156
Maryland	2
Massachusetts	1189
Michigan	409
Minnesota	1372
Mississippi	729
Missouri	1090
Montana	2433
Nebraska	40
Nevada	770
New Hampshire	252
New Jersey	1981
New Mexico	1018
New York	192
North Carolina	2364
North Dakota	651
Ohio	649
Oklahoma	1241

(continued)

Table 4.1 (continued)

Oregon	2215
Pennsylvania	1488
Puerto Rico	0
Rhode Island	113
South Carolina	465
South Dakota	852
Tennessee	391
Texas	6713
Utah	1493
Vermont	96
Virginia	410
Washington	1646
West Virginia	1230
Wisconsin	781
Wyoming	828
United States	**59,362**

Top 10 states at high to extreme wildfire risk in 2019
Source: National interagency fire center

Again, I emphasize that such response-related costs are largely the result of human behavior and, therefore, reflect decisions made (or not made) by governmental authorities who are necessarily respondent to their baseline popular-support. It is reasonable, therefore, to seriously consider whether it may, nonetheless, be cheaper to *remove (or at least ameliorate) the major precursors of a wildfire than to pay both the monetary and human costs of simply responding to a wildfire once it happens?* I will pay more attention to this issue regarding *precursors to wildfires* later in this chapter .... as I will in all of the following chapters in Part II. But first, I focus on the *interconnection of causal chains* associated with wildfires.

Figure 4.3 depicts a number of these interconnections. I caution the reader to consider that these causal chains are based largely on our collective experience and should not be viewed as complete; they are simply examples that, taken together, inform a holistic view of how emergency planning and response can and (I suggest) should be conceptualized in the twenty-first Century. I use subheadings for the following discussions of Fig. 4.3 to facilitate the reader's keeping track of key issues.

4.1 Loss of Wildlife Habitat

Humans are not the only animals to run away from a wildfire; so do deer, cows, mountain lions, as well as insects and other biotic vectors, including those that can physically injure or infect humans, whether they be first-responders or residents.

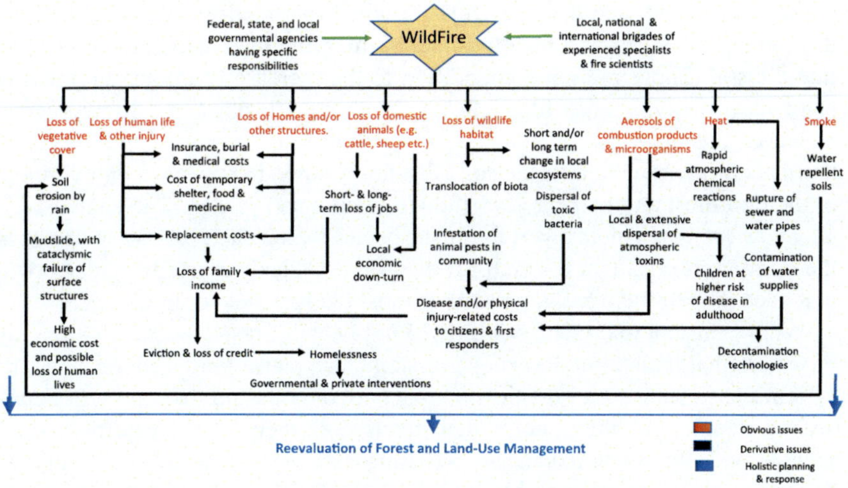

Fig. 4.3 Interconnections of causal chains associated with wildfires

Certainly, any response effort in a wildfire includes immediate on-the-scene medical first aid. However, some physical injuries and all infections require hospital-level care that can take time and is always expensive, with the too frequent consequence of job-loss or, otherwise, severely lowered family income. Keep in mind that even as recent as 1918, well over 25 million Americans did not have any health insurance.

Depending on the type of burned vegetation, soil changes caused by the fire, and the density of viable seeds, it can take months and even years (if ever) for the pre-burn ecosystem to recover. Given the extent of the burn, as well as the frequency of subsequent burns, opportunistic plant associations may establish themselves, thus providing suitable habitats for animal communities quite different from previous communities. We humans tend to view the so called natural environment as an unchanging 'scenery', but wildlife biologists, botanists, and woodland and grassland ecologists see that same environment as a dynamic system that evolves over time, sometimes slowly, sometimes rapidly, depending on a variable collage of factors. While any ecosystem certainly presents a 'scenery', its dynamics derive from its being a home-base of intertwined competition, mutualism, predation, reproduction, and morbidity among the smallest bacterium up to the largest carnivore and tree.

A good (but, also extreme) example of how an ecosystem can change abruptly is the impact of slash-and-burn agriculture in a tropical rainforest where annual rainfall is on the order of 200 inches and annual temperatures range on the order of 80–95 °F. Few places in the United States have the appropriate climate for world-class rainforests, but certain areas of Hawaii, North Carolina, Washington, Alaska, and California come somewhat close.

Basically, slash-and-burn agriculture, which is practiced world-wide, involves the cutting down of rainforest trees and burning the logs, which releases nutrients into the soil. Those nutrients in addition to the increased sunlight admitted to the erstwhile canopied forest-floor allows the farmer to plant and grow vegetables. However, this only last for a couple of years because the heavy rainfall washes the nutrients away, leaving a sterile soil. The farmer then has to move on to other areas of the rainforest and make a new (though temporary) garden. The abandoned (and now nutrient-poor) plots revert to brush species which can grow on sparse nutrient; the original trees cannot re-establish themselves. The ecosystem changes: wildlife that once thrived there are gone and are most likely never coming back.

What's this got to do with wildfires? Simply, wildfires, depending on the myriad of factors that maintain a forest or grassland, can cause long-term changes in ecosystem dynamics having significant impact not only the animals and vegetation, but also on people …... their health, their livelihood, their lives. Keep this thought in mind, especially as you read Chap. 9 (Climate Change). After all, my brief description of the effects of slash-and-burn agriculture depended on climates characterized by high temperature and heavy rainfall, factors that might very well cause Americans to re-evaluate their understanding of the *combined effects of wildfires and climate change*.

Well, let me turn your attention to another aspect of losses associated with the burnout of habitat…..but, this time, the focus is not on wildlife and vegetation but, rather, on those who first respond to wildfires. They have lives too. They also have to experience the risks of wildfires. They are the only living beings who don't run aware from a fire but run to it.

Of the estimated 1,115,000 firefighters in the U.S. in 2018, 33% were career firefighters and 67% were volunteer firefighters (Ebarts & Stein, 2020). It was not until 2012 that the U.S. Office of Personnel (OPM) issued an interim final rule, under which "wildland firefighters and fire protection personnel will be eligible to obtain health insurance for themselves and their families under the FEHBP (Federal Employees Health Benefits Program)." The rule also permits agencies to request that OPM authorize them to offer FEHBP coverage to other workers who perform emergency response services similar to the wildland firefighters" (Chief Human Capital Officers Council, 2012).

Good for wildfire-fighters; about time. But that still leaves many volunteer and career non-wildfire fighters, as well as first-victim, non-firefighter victims in insurance-limbo.

Regarding volunteer firefighters, consider the following quote:

> Firefighters are subjected to rigorous physical exertions in the line duty. Some responders are required to complete an initial examination to determine their ability to do the job; however, *many volunteer departments do not require follow-up annual of biannual testing* to ensure their members stay fit over the course of their tenure. In a 2014 independent study conducted by NVFC (National Volunteer Fire Service), *only 18 percent of volunteer respondents said that their departments required an annual health* screening. NVFC has adopted a position that supports annual medical assessments. (U.S. Fire Administration, 2016: italicized emphasis added by author).

Thusly do firefighters who do not receive appropriate medical monitoring disproportionately suffer not only heart attack, stroke, and respiratory disease, but also an increased probability of developing a wide range of cancers.

4.2 Loss of Domestic Animals

The wildfire loss of domestic grazing animals (cows, sheep, horses, goats, etc.) are not only devasting to these animals but also, obviously, to their owners who are financially dependent on them. The cost of raising a cow from birth to weaning is variable due to several factors, including the cost of corn fed to the calf and the number and quality of available graze-acres.

Counting in other costs, such as medicines, heat and electricity, etc., it takes somewhere between 500 and 1000 dollars to raise a cow; let's call it, on average, 800 dollars. If you're raising 400 cows, that comes to 320,000 dollars.

I choose the number 400 because I've recently read an article by a rancher who had 400 animals (cows and calves) in a grazing-range decimated by a wildfire that finally burned through 319,000 acres in California. He was able to retrieve 100 still alive, but some were badly scalded; he expected over half of those would not survive or must be euthanized. Given his devastation of his losses, he began to advocate land management practices … had to be changed more in keeping with the way the First Peoples managed their land for over 13,000 years. He reminisced about how his grandfather was once burning some brush and was accosted by a crew from the state Department of Forestry and Fire Prevention who extinguished his brush fire and proceeded to sternly lecture him against ever doing it again. So much for paying attention to old knowledge that works (David Daley, 2020).!

4.3 Aerosols of Combustion, Heat and Smoke

Wildfires, like any fire, emit a variety of chemical compounds, including carbon dioxide (CO_2), carbon monoxide (CO), and particulate matter (e.g., white and black carbon). Carbon dioxide is an asphyxiant (i.e., displaces oxygen in any confined space); carbon monoxide is a poison. Depending on the size of particulates, the larger particles may cause burning eyes and a runny nose; exposure to the very small particles (referred to as $PM_{2.5}$; particulate matter 2.5 microns in diameter, about 1/30 the diameter of human hair) can result in heart and lung diseases (including cancer) and, of course, premature death.

Depending on the type of vegetation burned and flaming-temperature, wildfires produce additional combustion-products, including a variety of volatile organics (VOC_s) such as benzene, which is a known carcinogen. High temperature-burns approach a state of complete combustion with lots of black smoke; lower temperature-burns result in increasingly lower degrees of combustion and produce white smoke.... the colors reflecting different chemical constituents ranging from slight irritants to toxins.

Temperature not only directly influences which chemical aerosols are produced but also speeds up the rate at which those chemicals interact with chemicals (e.g., vehicular exhaust) already in the ambient air. Such chemical interactions can therefore result in toxic clouds that can spread far and wide from the wildfire site. This phenomenon is a good example of how a local wildfire not only affects the health and well-being of local wildlife and humans, but also significantly distant populations.... populations that typically don't fall within the categories of 'responder-victims' and 'resident-victims' but rather, let us say, in the context of earlier discussions, 'distant strangers'.

I think it important that I add here a comment about such 'distant strangers' because such a category includes, of course, children who, having been exposed to atmospheric toxins (e.g., carbon monoxide, volatile organic compounds, nitrogen oxide), are at a higher risk of various diseases (e.g., cardiovascular disease, depression, attention deficit disorder) in adulthood (CDC, National Environment Public Health Tracking; Population Reference Bureau, n.d.).

In Chap. 3, I referred to such victims as 'generational victims', victims who suffer the effects of an emergency well after the bells and whistles have become silent, TV-coverage ended, and the emergency long-since declared 'Over'. No, for them, it's not.

Do we really think we can draw a neat line around the consequences of a wildfire? Is a singular emergency (and our response to it) to be circumscribed by not only a time-limited response effort but also a sharp geographic boundary?

Now, let's turn to the issue of smoke from a different perspective .... one that is recently receiving a lot of attention by the scientific community. Based on some preliminary experiments with wildfire-smoke and soils, data indicate smoke (whether hot or cold) may increase water repellency. The experiments were originally conducted on sand-soils, but future experiments are planned using soils typically characteristic of wildfire-areas (Desert Research Institute, 2021).... a good example of how scientific understanding depends on ever-inventive experimentation.

I have therefore included a line-of-causation between 'smoke' and 'soil erosion' in Fig. 4.3 as a possibility well-worth considering in planning emergency response actions. After all, the loss of vegetative cover is a known precursor to soil erosion and, as I have previously discussed in Part I, to subsequent mudslides (often referred to as 'post-fire debris-slides' with high economic costs and posible loss of human lives.

Depending on type of vegetation, soil-nutrients, density of viable seeds, weather and, of course, the frequency of fires, it can take years and even decades for a burned-out area to revegetate. If, in fact, wildfire-smoke does result in an increased repellency to water, then erosion and its consequences will be speeded-up.

While mudflows are typically associated with heavy rainfall, even relatively minor rains can result in flashfloods and subsequent mud-debris flows. The USGS (U.S. Geological Survey) provides free post-fire debris-flow Hazard Assessments for recently burned areas in western states to any interested Federal, State, or local agency, as well as to any private organization having appropriate concerns.

4.4 Change in Ecosystems

A second consequence of prolonged revegetation is the prolongation of ecosystem-recovery or ecological-transformation. Yes, ecosystems change, and sometimes, very dramatically. Perhaps among the most profound (and among the most ancient examples) is the Amazon.

About 60 million years ago, a very large asteroid slammed into earth close to the north-eastern edge of South America and triggered a major extinction-event. The dinosaurs disappeared. But the asteroid impact that caused the demise of dinosaurs also gave rise to what we today call tropical rainforests which ultimately spawned a much larger diversity of both plant and animal species than previously existed (Duong, 2021). Asteroids can obviously bring about drastic ecological change. But humans can do the same.

Considering the impact of human pollution from the late 1700s through today humans have had more impact on ecosystems than the last glaciers (Pennisi, 2020).

4.5 Loss of Human Life, Homes & Other Structures

Beyond the psychological trauma regarding the loss of lives and homes, there is always the problem of finances. As I have said previously, money counts. Whether first- responder-victim or residential-victim, the costs associated with health insurance, burial & medical care, temporary shelter, food & medicines, and the replacement or refurbishment of home or business.... all these needs compete with one another, especially regarding their dependency on a viable job. Without that job, which for the vast majority, is the primary source of money, survival becomes an open question as opposed to a given.

Depending on the extent of wildlife losses and prevalent local employment (e.g., agriculture, tourism, and other seasonal job-opportunities), it is difficult to identify an overall pattern to the economic consequences of wildfires among different communities.

Differences between the (even temporary) increase of certain employment opportunities presented by fire-suppression efforts (e.g., selective clearing of brush, provisioning rest-areas for firefighters, etc.) and, on the other hand, the loss of employment due to fire-related depression of local commercial business (e.g., restaurants, bed &breakfast rentals, etc.) contradict each other (Moseley et al., 2013).

However, it is clearly evident that local, seasonable labor (e.g., related to tourism and the availability of natural resources (e.g., lumber, scenic views, campgrounds, etc.) is typically and severely curtailed. Likewise, small businesses that otherwise provide substantial full and part-time employment opportunities may be burned-out of existence or otherwise fail due to fire-mediated and/or fire-enhanced illnesses (e.g., respiratory illness, microbial and parasitic infection etc.) and drastic loss of market (Guary & Davis, 2020).

Such differences in simultaneous, economic gains and losses result in substantive inequities in individual family-incomes with subsequent consequences with respect to the unequal distribution of the manifold effects of even short-term financial instability, not to mention immediate and long-term personal disaster, including depletion of savings, loss of credit, and possible eviction. Keep in mind that well over 50% of Americans live paycheck-to pay-check and, even in the absence of wildfires or other disasters, live in constant economic peril.

4.6 Overview of Fig. 4.3: Holistic Response

This figure (as well as those in following chapters) is my attempt to depict a wildfire as not a single event, but as a series of interconnected phenomena leading to and deriving from fire. Historically, the response to a wildfire costs many millions-to-billions of dollars, which raises a number of questions:

- Whether that money should be better spent preventing the fire in the first place?
- What political and private-sector decisions regarding land-use control and wilderness-management might be made to minimize the probability of wildfires?
- Is there a time- and geographic limit to the diverse social, economic, and environmental sequalae of a wildfire?
- What are alternative ameliorative procedures for aiding and assisting victims over extended periods of time?
- How do we respond to the various inequities associated with wildfires?
- Is it appropriate (if not advisable) to reconsider how emergency-planning (as opposed to emergency-response) is conducted to ensure the critical involvement of the natural and social sciences, as well as other social- economic (public and private) services? (Fig. 4.4)

The above questions are premised on what I refer to as a "holistic view of emergency planning and response". I use the term 'holistic' in precisely the way it is defined in the Oxford American Dictionary (3rd edition):

> The theory that parts of a whole are in intimate interconnection, such that they cannot exist independently of the whole, or cannot be understood without reference the whole, which is thus regarded as greater than the sum of its parts.

It is in this sense that a wildfire (as well as any emergency) is not a singular event, but rather a single page among many interconnected events, including precursor and subsequent events.

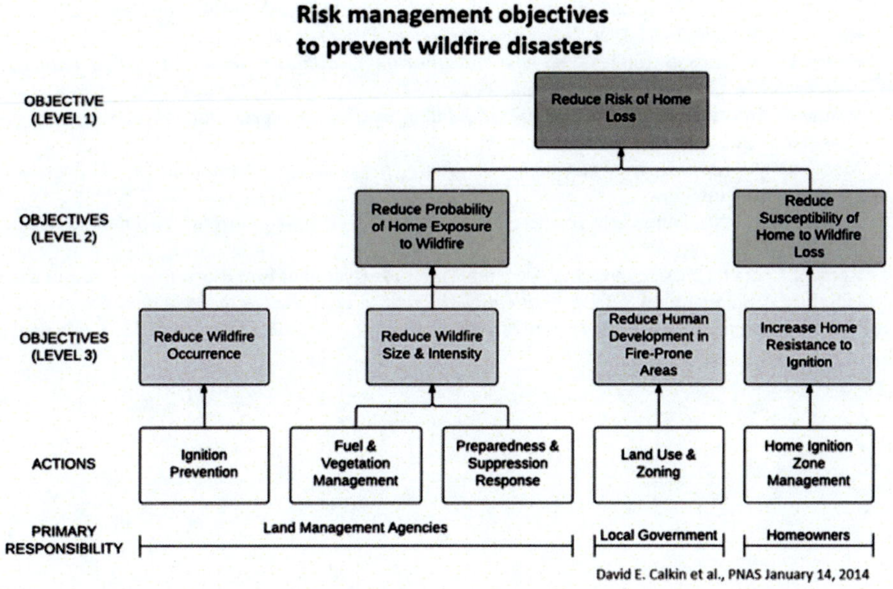

Fig. 4.4 Risk management objectives to prevent wildfire disasters. (Calkin et al., PNAS January 14, 2014)

Suggested Exercises for This Chapter

Use the Internet to look up recent (i.e., within past 5 years) wildfires in California. Choose one for which you can identify (even by rough estimate) the total coast of response efforts. Considering that figure, as well as the various wildfire-related consequences, respond to any one (or more) of the following 4 exercises.

1. Discuss the reasonableness of paying for response efforts rather than paying for efforts that focus on the prevention of the wildfire in the first place.
2. Discuss in detail the practical efforts to be taken to minimize risks due to animal and infective vectors related to the migration of wildlife into residential areas.
3. Identify and discuss how one might ameliorate the health risks of wildfire smoke and aerosols to humans within the fire-zone and at a far distance.
4. What steps might reasonably be taken to ameliorate the inequitable loss of employment among populations directly impacted by wildfire.

References

Calkin, D., et al. (2014). PNAS (Proceedings of the National Academy of Science of the U.S.A).
CDC (Centers for Disease Control and Protection). National Environment Public Health Tracking.
Congressional Research Service. (2021).
Daley, D. (2020, September 27). *I cry for the mountains*. Enterprise-Record.
Desert Research Institute. (2021, May 25). *Does cold wildfire smoke contribute to water repellent soils in burned areas?*

Duong, F. (2021). Dinosaur-killing asteroid probably created the Amazon rainforest. *SCIENCE.*
Ebarts, B., & Stein, G. (2020). *U.S. Fire Department Profile 2018.*
Guary, D., & Davis, S. (2020). *Full cost of California's wildfires to the U.S. revealed, nature sustainability*. University of California, Irvine.
Insurance Information Institute. (2021). *Annual number of acres burned in wildland fires, 1980–2020.*
Moseley, C. (2013). *The economics of large fires.* Digital Commons as University of Nebraska-Lincoln.
Pennisi, E. (2020). Humans have altered North America's ecosystems more than melting glaciers. *SCIENCE.*
Roman, J., et al. (2020, November–December). NFPA (National Fire Protection Association).
U.S. Fire Administration. (2016). *Critical health and safety issues in the Volunteer Fire Service.*
U.S. Fire Administration. (2021). *Fire estimate summary, April 2021: Residential building cooking fire trends.*

Chapter 5
Pandemic

The World Health Organization (WHO) declared Covid 19 a pandemic on March 11, 2020. As I write these words (15 months later) Covid 19 has killed more than 600,000 Americans and, world-wide, roughly 4 million. It is difficult for most of us to deal with such large numbers. I therefore offer a telling (though grizzly) context: the total number of U.S. soldiers killed in World War 2 plus those in Korea plus those in Vietnam was 504,224; Covid 19 has killed almost 96,000 more Americans in 15 months than those cumulative wars did over a good number of years.

The last time that America experienced a pandemic was in 1918–1919. During that pandemic, between 40 and 50 million died world-wide; about 675,000 in the United States. Well, that was over a hundred years ago (Oldstone, 2010). Today, suddenly faced with Covid 19, Americans tend to think pandemics are brand new. They are not.

Pandemics are infections that are spread world-wide, but it is only in relatively recent times that we have acquired a fully global perspective; for the most of human history we had only very local experiences. Up until Columbus, the world (as we know it) was significantly smaller; North and South America, though inhabited, was not known to western civilization. A couple of thousand years prior to Columbus, the world was very tiny to the average Roman of Greek. Extensive travel then was, at best difficult, tedious and especially dangerous due to two omnipresent factors … warriors and diseases that had become established in all the major civilized cities since 500 BC (McNeill, 1977, 1989, 1998).

The word 'epidemic' means an infection that is spread within a defined geographic area which could be a city, a region, or even a whole nation. How does an epidemic become a pandemic? That's easy: by the frequency and extent of human travel throughout the globe. Micro- and macro-parasites (e.g., worms, etc.) go along for the ride within their hosts. As of today, one can travel halfway around the world in 22 h. (counting 1 or 2 stops); this means that, depending on the direction you take, you can go anywhere in the world with 22 h. My bet is that, within 10 years, given our technological cleverness, that travel-time will be cut in half.

P. A. Erickson, *Effective Environmental Emergency Responses*,
https://doi.org/10.1007/978-3-031-05893-6_5

Taking advantage of the ease of modern (and future ultra-modern) travel, many thousands of Americans and comparable numbers of foreign nationals travel back and forth throughout the globe (Fig. 5.1) … a globe filled with known and unknown micro- and macro-parasites. Who get's infected? Who spreads that infection? Almost impossible to tell! It might be an American who travels to France and contracts an infection from a Belgian who just returned from a quick visit to South Africa. Who knows?

Consider our experience with HIV [Human immunodeficiency virus; generally referred to as AIDS (acquired immune deficiency syndrome)]. First identified in 1981, HIV has since killed somewhere between 50 and 100 million worldwide, and over 700,000 in the U.S.; about 13,000 Americans die each year.

Early cases in America were among homosexual males; mass-media therefore quickly dubbed the infection a 'homosexual disease' which, of course resulted in a rabid increase in the already national recrimination of gays. The subsequent identification of a gay male-airline steward who had deplaned in Africa and was later diagnosed as HIV-positive became the 'fall-guy' for the introduction of HIV in American. Known as 'Patient Zero', he died of complications with HIV/AIDS in 1984 with that sobriquet which, as it turned out, was utterly wrong.

It took over 30 years and the use of newly invented biochemical, genetic, and immunological technologies to demonstrate that HIV migrated from Africa to the Caribbean and thence from Haiti to New York and, finally to different places in the U.S. including California … a translocation of HIV that occurred well before HIV

Fig. 5.1 American and foreign national travel

was first diagnosed in Los Angles (and therefore in the purported 'Patient Zero' (Akpan, 2016).

The fate of 'Patient Zero' (the first confirmed case) is a good example of how preexisting prejudices (i.e., male homosexuality) and biases, when coupled with fear and panic, and lubricated with infantile logic are always incompatible with scientific understanding. Moreover, they become the breeding grounds for misinformation, outright lies, and self-serving excuses for decisions made on biased logic and not on scientifically demonstrated fact.

The frequency of pandemics is today largely influenced by the frequency of global human-travel with millions to and fro from the U.S. every year (Crosby, 2003). For many millions of years, global dust and sandstorms were the primary vehicles of pandemic diseases; today. Human travel is the primary vehicle.

Obviously, we can't stop people from traveling, nor can we stop winds from blowing. Pandemics are intrinsic phenomena of our extremely dynamic planet.

Before proceeding with a holistic review of our current Covid 19 pandemic, I think it appropriate to consider the 1918 influenza pandemic. The two-year global death toll of this pandemic was roughly 50 million. In America, the toll was 675 thousand.

At that time, we had no knowledge of viruses; that word did not exist. All we knew was that there was something very much smaller than bacteria existed and might be causing the then unknown disease. There were no "authoritatively approved" vaccine for the that disease; what passed for what we now call vaccines were homemade concoctions made by individuals who knew nothing about the disease.

Which leads me to suggest you serious consider the following four suggestions regarding pandemics in general and, of course, specifically the Covid 19 pandemic as we have thus far experienced it.

1. If you ae going to travel (anywhere outside of your home-locality), consider first seeing a physician who specializes in 'travelers medicine' (the technical term is 'emporiatric medicine'). Tell that physician where you're going so that she/he can determine what medicinals are most appropriate and what foods and drinks you should avoid. By the way, even if you're traveling in the United States, you are well advised to make yourself aware of diverse health-risks that vary from State to State as well as within a single State.
2. Stop this nonsense regarding the naming any disease after a Nation or an ethnic group. It makes no scientific sense, but only satisfies an agonistic and xenophobic proclivity based on ignorance of epidemiological facts.
3. Don't think that any of us will be free of future pandemics. Global travel is only going to increase in the years to come. Travelers whose immune systems reflect disease-threats within their long-term residency easily become infected in distant climes and, as easily, bring both old and brand-new diseases back home. And, of course, we can't stop the winds from blowing.
4. Consider: If you have a heart attack, you go to a cardiologist; if a brain tumor, to a neurosurgeon, if an auto-malfunction, to a licensed mechanic. By the same

logic, if you are faced with an epi- or pandemic, you should go to a physician that specializes in infectious diseases … not a cardiologist, not a neurosurgeon, not to a mechanic, and may I add, certainly not to a politician, or a TV talk-host, or some sport-celebrity. And don't be confused by academic titles, such as 'Dr.' and 'Ph.D'. Dr. in what? In dentistry? Ph.D. in what? In podiatry? No, you go to a fully accredited and seasoned physician well trained in infectious diseases whatever her of his honorific title may be.

5.1 World Pandemic, American Syndemic

Figure 5.2 is an overview of selected interconnected causal chains that are well documented in the American experience of the Covid 19 pandemic. But perhaps a better phrase is 'Covid 19 syndemic', the later word used by epidemiologist and medical anthropologists to denote health-disparities driven by and/or exacerbated by political and social inequities.

It has been argued that, while here is a global Covid 19 pandemic, the American experience with Covid 19 is very clearly syndemic in the sense that preexisting conditions, racism, mistrust in science and leaderships, and a fragmented healthcare system have resulted in its rapid spread throughout populations and have caused more death and devastation than any other factor (Horton, 2020)."

As indicated under the obvious consequences (i.e., Deaths and Loss of Jobs) of Covid 19, there are obvious statistical differences in death and death rates based on

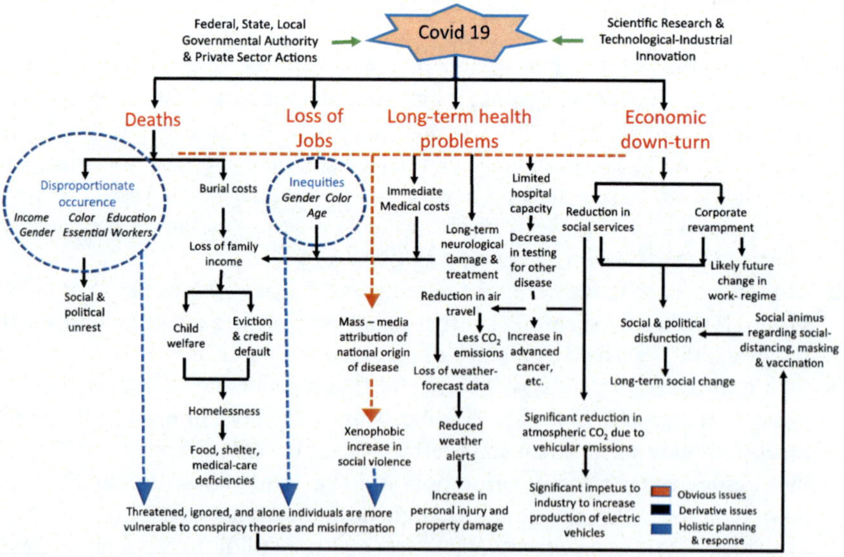

Fig. 5.2 Interconnections of causal chains associated with covid pandemic

income, color, education, and gender as well as differences in the number of jobs lost based on color, gender and age.

In a recent epidemiological study, the authors conclude that the prevalence and severity of the Covid-19 are influenced by *individual non-communicable diseases and social conditions* (Bambra et al., 2020).

5.2 Deaths & Loss of Jobs

Because gender is a common discriminator related to both death and job-loss in the American Covid syndemic, and because it is a major excuse for social inequities (Fig. 5.3) it deserves closer examination.

Might women have a higher health-risk because they are typically paid less (even for the same work) than men and, maybe, are not offered or, maybe, cannot afford health insurance? Or, maybe, the fact that she makes less money *and the fact* that children are being home-schooled and therefore require adult-supervision *and the fact* that, "as everyone knows, a woman's place is in the home" collectively interact to generate her pink-slip?

And what about those women who are single parents or those who do not have children?

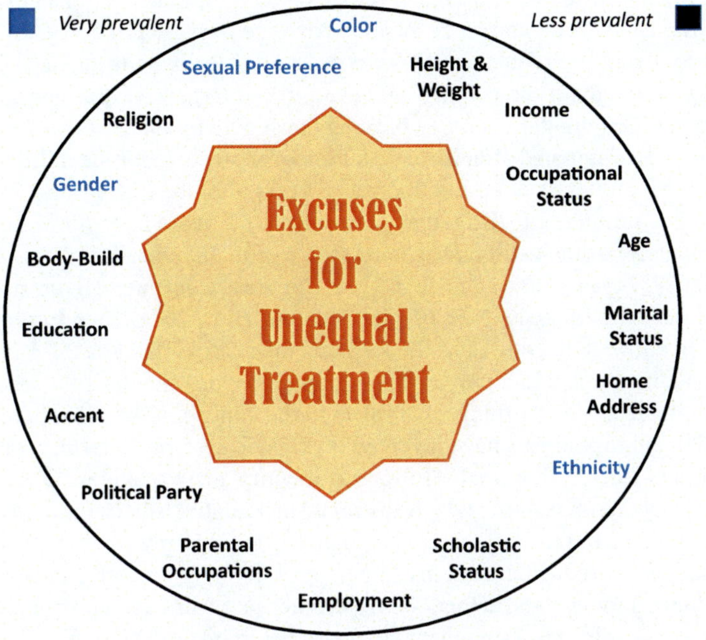

Fig. 5.3 Common discriminators of social inequities regarding covid pandemic

Consider, also that, in 2019, same-sex married couples numbered 543,000, while same-sex unmarried couples numbered 469,000 (U.S. Census Bureau, 2019).

By the way, while homosexuality is typically viewed as an aberrant behavior in humans (if not an outright sin) and deserving of disgust and social approbation, it is rather common in the animal kingdom, documented (so far) in at least 1500 species, including bonobo chimpanzees, lions, dolphins and killer whales, geese and ducks, black-headed gulls, giraffes, crabs and worms, and mammals that have complex herd-lives (News: Medical-Life Sciences, 2006). Meanwhile, humans continue to torture and murder and otherwise persecute fellow humans as committing sexual acts considered as being against Nature when, in fact, homosexuality is actually Nature's invention (Rutherford, 2019). Bigotry is a human invention that is based on ignorance reinforced by the human proclivity to present one's own behavior as rightfully the standard for the behavior of everyone.

Maybe gender-based statuses of humans within what we call a 'family' is becoming somewhat of a social-dinosaur. Maybe he words 'husband' and 'wife', which are obviously gender-based and of long-historical usage are giving way to the neuter-noun 'partner'. After all, over the last 50 years, the heterosexual-divorce rate in America has varied somewhere between 40% and 50% … hardly a statistic that verifies the concept of marriage as a preeminent characterization of the American social system. It would seem that our use of the word 'gender' and its various associations with interpersonal relationships (e.g., marriage, sexuality, and individual competence) are more indicative of the persistence of archaic means of social marginalization rather than social solidarity.

In 1942, because of a shortage of pilots, the U.S. military set up an experimental group that came to be known as Women Airforce Service Pilots (WASP). It was a tremendous success, with over 1000 women-volunteers who tested new planes, ferried them throughout the country, and even towed targets for training ground- and air-gunners. Despite those women proving they could fly planes as well as men, that program was terminated after 2 years. It took over 30 years for the military to finally grant WASP military status, and that was 6 years after the first woman in American history was promoted to Brigadier General and, 2 years later, to Major General. Unfortunately, women still have a hard road to equality ahead of them. After all, it has been 232 years since the first U.S. Congress convened. Even now (2021), women account for only 24% of our Senators and 27% of our Representatives. "Well, ya know, they gotta raise the kids and take care o' the house." Buls…!"

Having mentioned in a previous paragraph the word 'family', the teacher-in-me requires me to say a few things that relate to the concept of social solidarity.

Social anthropologists have invented 6 "kinship systems" … rules which different cultures (in the thousands) follow to identify their relatives. One of these is called "Hawaiian Kinship", which distinguishes relative solely on the basis of sex and generation. In this system, every related woman in my mother's generation is "Mother"; every related man in my fathers generation is "Father", and every sibling of my many mothers and fathers is my brother or sister (Schwimmer, 2001). This system is quite different from that known as "Eskimo Kinship", which is commonly used in the U.S. and whereby I have only one mother and one father; the rest of related males and females are uncles and aunts and I call their all their children cousins. My own siblings I call brothers or sisters.

I think it very clear that the Hawaiian system promotes a bonding among relatives while the Eskimo system promotes isolation. I'm not suggesting we change whatever kinship system we have. But I am suggesting, as a Nation comprised of both Indigenous- and Immigrant-derived populations, we do have to reassess those ideas and concepts we are born to and strive, rather, to implement a more fair social solidarity in place of the current inequitable isolations of our fellow humans.

To accomplish this, yet another word that should be consigned to oblivion is 'race'. When we Americans use that word, we most often mean 'color'. In fact, early social-anthropologists carried color-charts with them on their various global travels which they placed under the armpits of amused indigenous humans to define precisely a particular skin-color unaffected by sunburn. Depending on the charted color, people were defined as belonging to a particular 'race'. It is far too late for use to ask those students of humanity why they did not choose eye-color, or hair-color, or length of earlobe, or penis-length, or … whatever as the proper badge of humanity. None of these characteristics tell us really anything of significance, except, perhaps the migratory path of ancient ancestors.

No, there is no modern biological meaning to the 'race'. All humans are genetic mutts as a result of tortuous lines of sexual unions extending well beyond 200,000 years which, by the way, is equivalent to well over 8000 generations. Nonetheless, using 'color' as a workday-synonym of 'race', and mixing it with 'ethnicity', we see (as in Table 5.1) varied inequities among these most-often arbitrary groupings regarding numbers of Covid 19 cases, hospitalizations, and deaths. The question is "Why are there such inequities?" Despite the prolific agonies of Covid 19, they at least underscore and make clearly visible the necessity for reassessing any social structure and function that promotes marginalization of anyone.

Maybe, just maybe the twenty-first Century will become known as the time Americans began to stop thinking in terms of archaic words and categories that have too long persisted beyond any meaningful contemporary utility other than

Table 5.1 Risk for COVID-19 infection, hospitalization, and death By race/ethnicity

Race and ethnicity are risk markers for other underlying conditions that impact health

Rate ratios compared to White, Non-Hispanic persons	American Indian or Alaska Native, Non-Hispanic persons	Asian, Non-Hispanic persons	Black or African American, Non- -Hispanic persons	Hispanic or Latino persons
Cases	1.6×	0.7×	1.1×	2.0×
Hospitalization	3.3×	1.0×	2.9×	2.8×
Death	2.4×	1.0×	1.9×	2.3×

Race and ethnicity are risk markers for other underlying conditions that affect health, including socioeconomic status, access to health care, and exposure to the virus related to occupation, e.g., frontline, essential, and critical infrastructure workers

Content source: National Center for Immunization and Respiratory Diseases (NCIRD), Division of Viral Diseases prejudice, bigotry, and outright malicious disparagement punctuated with hate and agonistic behavior.

5.3 Economic Down-Turn

Figure 5.2 shows that Covid 19-mediated deaths and job-losses together contribute substantially to that pool of threatened, ignored, or otherwise isolated citizens who become more vulnerable to the wide and seemingly constant dissemination of conspiracy theories, misinformation, and outright lies (Shute, 2021). Included in this pool are also those who, not marginalized by social-dynamics, nonetheless individually and collectively experience the anxiety, fear and even anger always associated with living day-to-day in the midst of a rampant killer- or otherwise debilitating disease along with its implications regarding personal income. Under such circumstances, it is not surprising that many decisions are made that are not appropriate (even not rational). Anxiety and fear alone are sufficient to drive anyone into a protective mode of focused self-survival. Add the spices of anger, confusion, the loneliness imposed by personal-quarantine, the avoidance of crowds, social-distancing even in small groups, and the bedlam of voluminous and contrary pundits' opinions and advice, the result is a social animus, the full consequences of which cannot be foretold with any certainty. Of course, a few consequences are clearly discernable regarding Covid 19, including the refusal to wear masks or to employ social distancing (common behaviors in the 1918–1919 American pandemic), and their persistent (though often premature) insistence to 'return to normal'.

Two obvious consequences of a Covid 19-driven decline in economic health include, of course, a reduction in social services and what I refer to as corporate remampment; both are direct consequences of social-distancing (including, at the extreme, quarantine) requirements imposed as very effective measures to minimize viral-spread.

While such requirements are absolutely necessary from an epidemiological perspective, they ae also anathema to Americans' daylily familial, neighborly, and job-related social interactions and, thereby exasperate not only their frustration, but also their anger.

It is therefore not surprising that social distancing, fear of contracting disease, and economic uncertainty have combed to increase the percentage of American adults who have reported symptoms of psychological distress from 3.7% in 2018 to 24% in 2020 (McGinty et al, 2020).

We will have to wait to see any long-term effects of these personal and social disfunctions; no doubt, however, that there is a significant probability that there will be (at least, some) tangible, long-term social changes, especially regarding social inequities in general.

Meanwhile, it is becoming evident that industry, already primed by the advent of global-communication and digital-integration of the diverse elements of 'supply chain management', is more than just tinkering with Covid19-inspired development of not only school- but also workplace-virtual learning, production, and visual communication. I suspect that the latter technology will employ holographic hard- and soft-ware within a few years.

While we rightfully bemoan the many tragedies of Covid 19, American industry cannot be blind to certain positive consequences of virtual technologies, such as lower overhead costs associated with rental of office space, furniture, utilities, and (apparent) employee-preference to work at home with flexible schedules instead of spending otherwise fruitless hours commuting to and from an office run according to the office clock.

Another potential benefit to company-management is an apparent increase in the productivity of home-based workers when compare with office-based workers. I use the word 'apparent' because different studies have been done of different workers with mixed results. Even in cases were significant increases in productivity were noted, it cannot be assumed that such results would persist in post-pandemic times. Nonetheless, it is safe to say that enhanced productivity by at-home, remote workers may be real (if not necessarily universal). Certainly, it is worth exploring.

For now, it is expected that about a quarter of the American workforce will work remotely through 2021 (Apollo Technical, 2021). We just have to wait and see what eventually evolves in the American workplace in post-pandemic times.

An actual positive consequence of the reduction of social services (as one measure of economic down-turn) is the significant reduction in atmospheric CO_2 (a greenhouse gas) and other exhaust-wastes in both ground- and air-transport emissions. With respect to 'Climate Change', these reductions are universally seen as positive. Regarding ground-vehicles, such reductions are commensurate with the on-going transition from carbon-based fuel to electrical energy; regarding aircraft, they reinforce political pressures on the aircraft industry to redesign engines and fuels to decrease overall combustion products. In this sense, the negative impacts of Covid 19 on air- and ground- travel may be viewed as having positive impacts on air-quality and, thence, on human health as well as climate change. But, of course, this depends on decisions made (or not made) by humans, that uniquely spurious species which is so infuriatingly difficult to predict.

5.4 Immediate & Long-Term Medical Costs; Limited Hospital Capacity

Immediate and especially long-term medical costs, when combined with a plethora of inequities (e.g., pay-rates, job-opportunities, gender, color, etc.,) can be absolutely devastating to individual- and familial-incomes, especially for those Americans (many millions) who live from paycheck-to-paycheck and who may also have lost family members.

Except for on-the-scene first-aid, first-responders do not have the time to worry about subsequent financial difficulties of first- or generational-victims and certainly do not have the wherewithal or the authority to ameliorate their financial problems. As we have seen during the Covid 19 pandemic, it is for politicians with governmental authority to act in this regard on behalf of all of us. Of course, whether they do so or not is, to say the least, problematical.

Congress passed 3 laws to provide some economic relief:

1. *The Coronavirus Aid, Relief, and Economic Security Act of 2020 (CARES Act)*
2. *The Covid-related Tax Relief Act of 2020 (Consolidated Appropriations Act)*
3. *The American Rescue Plan Act of 2021.*

Each of these acts provides payments to lower- and middle-income Americans. However, the IRS took 4 months to disburse 90% of allotted funds (Peter G. Peterson Foundation, 2021).

It is reported that on the order of 180,000 had closed at the very beginning of the pandemic; a slightly lower number (160,000+) within a few months (Sundaram, 2020). Hardest hit were small businesses (e.g., gift shops, breakfast and brunch cafes, men's clothing stores, flower shops, and other providers of personal services) employing thousands of low-income workers living from paycheck-to-paycheck. The 4 months delay in distributing CARES checks was unforgivable. How does one pay for rent, food, medicine, clothes, heat, electricity, water with no income for 4 months?

While seemingly to take a long time coming, the Consolidated Appropriations Act and the American Rescue Plan have provided significant relief, especially to small businesses, as have State and local governments, various non-profit organizations, and those landlords and creditors that have deferred rents and loans (Simon, 2021).

While we may allow ourselves to become somewhat optimistic regarding the National economy, there is one consequence of Covid that yet elicits despair. That is the fact that the number of Covid patients so overwhelmed hospital capacities as to preclude the normal screening and diagnosis of non-covid patients for other threatening diseases, including those (such as Cancer, High blood pressure or high cholesterol, Diabetes, Osteoporosis or weak bones, STDs (sexually transmitted diseases), Mental health conditions (like depression) which are increasingly difficult (if not impossible) to treat successfully if not detected early.

5.5 Special Consideration: Essential Personnel

Every emergency requires essential personnel who daily provide services that are absolutely necessary for maintaining the vital dynamics of social life. As we have experienced in the Covid 19 Pandemic (and, a hundred years ago, the 1918–1919 Flu pandemic) any pandemic affects every aspect of daily life, whether physical, psychological, social, political, economic, educational, spiritual, recreational, nutritional, medical … and on and on. Yet, amid the panic, confusion, and ever-present danger, there are those who continue to provide sustenance, care, and solace. They are heroes by any definition. As heroes, they are at high risk.

Such risks are various: contracting disease because they don't have the appropriate protective clothing or equipment; spreading disease to their family; having to come in contact with infected patients or with citizens not wearing masks and/or not practicing social distance; burnout from long hours of gruesome work; collecting curbside garbage that may be contaminated.

Nurses reusing PPE (personal protective equipment), canceling paid-time-off and working extended shifts are increasingly common workday practices that have resulted many to quit (Andrew, 2021).

Burnout is seen not only in nurses but across medical personnel, police, firefighters, emergency medical technicians and many other heroes who keep things running, including those who become designated as 'essential personnel' (e.g., delivery personnel, garbage collectors, etc.) who are required to daily expose themselves (and, therefore their families).

How does society intend to provide meaningful response in (terms of kind and duration) for these heroes while so many others publicly broadcast their bitter complaints against wearing masks and keeping social distance, as if such mandates curtailed their constitutional rights … as if their interpretation of those rights also granted them the right to infect their fellow citizens … as if their not being able to go to favorite restaurants or sport-events constituted a catastrophe greater than the pandemic … as if the inconvenience or discomfort of a mask was itself worse than a deadly disease?

How selfishly childish! Some of us are inured to selflessness, others, to selfishness. Both can evolve into habits. However, selfishness is a particularly difficult habit to break, probably because it is often a protective camouflage contrived of previous experience, or maybe a preoccupation with modern, easily available, and increasingly multitudinous means of achieving self-satisfaction. Whatever, it is always easy to say, "I feel so sorry for you" to someone who has experienced a harm. But how are you going to respond if that someone responds, "So what are you going to do about it?" That's the question that has to be answered by any comprehensive and effective behavioral response.

5.6 Overview of Fig. 5.2: Holistic Response

Because of the human propensity to travel and the persistent, microbially-contaminated sand and dust storms, we can do nothing to prevent the occurrence of a pandemic; we can only react to it. Given what we have personally experienced in the current Covid 19 pandemic and what we might have meanwhile familiarized ourselves with the 1918–1919 Flu pandemic, the following questions seem most appropriate.

- How do we propose to improve the integration of Federal and State efforts to achieve consistency regarding efforts to better control rates of infection?
- Should we consider alternative ways of stockpiling personal protective clothing and equipment to ensure ready access and equitable distribution?
- How do we deal with the problems associated with limited hospital-capacity?
- Does our experience with Covid 19 demonstrate a need to implement restrictions on mass media regarding the dissemination of outright lies and misinformation?
- Should mandates regarding masks, social distancing, quarantine, and vaccination originate from and/or be enforced by Federal, State and/or Municipal authority?
- What are alternative ameliorative procedures for aiding and assisting victims over an extended period?

Suggested Exercises for This Chapter

1. Using the Internet, select any single State in the U.S. and copy (or construct) a graph of Covid 19 infections from the beginning of the pandemic through the year 2002. Using that graph, identify and discuss actions taken (and not taken) by that State which may explain the graphical variances observed.
2. Explain in detail the consequences to America of not achieving a world-wide immunity to Covid 19 (or, for that matter, any virus).
3. Write an essay (giving specific examples) that discusses the probable and possible consequences of interstate travel in the U.S. during the Covid 19 pandemic.
4. Choose a debate-partner. One is to argue for the necessity of wearing masks during a pandemic; the other, to argue against the wearing of masks.
5. The Covid 19 pandemic has revealed serious inequities in the U.S. Select one that you consider most important and explain what Americans should do about it.

References

Akpan, N. (2016, October 26). America's HIV outbreak started in this city, 10 years before anyone noticed. *Science*.

Andrew, S. (2021, February 25). Traumatized and tired, nurses are quitting due to the pandemic. *CNN*.

Bambra, C., et al. (2020). The Covid-19 pandemic and health inequalities. *Journal of Epidemiology and Community Health*. https://doi.org/10.1136/jech-2020-214401

Crosby, A. (2003). *America's forgotten pandemic: The influenza of 1918*. Cambridge University Press.

Horton, R. (2020). Covid-19 is not a pandemic. *Lancet, 396*, 874.

McGinty, E., et al. (2020). Psychological distress and loneliness reported in 2018 and April 2020. *Journal of the American Medical Association*. https://doi.org/10.1001/jama.2020.9740

McNeill, W. (1977, 1989, 1998). *Plagues and peoples*. Random House/Anchor Books.

News: Medical-Life Sciences. (2006, October 23). *1,500 animal species practice homosexuality*.

Peter G. Peterson Foundation. (2021). *Note from Michael Peterson on the impact of the pandemic: How did Americans spend their stimulus checks and how did it affect the economy*. Policy Advice, 2021, Earthquake Statistics by Year.

Rutherford, A. (2019). *Humanimal: How Homo sapiens became nature's most paradoxical creature*. The Experiment, LLC.
Schwimmer, B. (2001). *Hawaiian Kin terms*. University of Manitoba (Updated: August 24, 2011).
Shute, N. (2021, January 24). When data shed light on societal challenges. *Science News (editorial)*.
Simon, R. (2021, April 16). Covid 19's toll on U.S. business: 200,000 extra closures in pandemic's first year. *The Wall Street Journal*.
Sundaram, A. (2020). Yelp data shows 60% of business closures due to the Coronavirus pandemic are now permanent. *CNBC*.
U.S. Census Bureau. (2019). Release number CB19-TPS.51.

Chapter 6
Hurricane

A simple recipe for a hurricane:

1. Prolonged sunlight + seawater = warm, moist air that rises, which causes low pressure beneath
2. Air-rush into low pressure + spinning earth (Coriolis effect) = rotation of Wind (counter- clockwise north of the equator; clockwise, to the south).
3. Condensation of moisture in upper atmosphere = increased power of circulating wind.

Voila! A hurricane, or as they call it in the west, a typhoon or cyclone.

Given the role of ocean-temperature in forming hurricanes, we can define a seasonality to them (for Southern and Eastern U.S. Coasts, June-November 39), though, given the possibility (or probability) of Climate Change (Chap. 9), that time-frame may have to be changed.

Excepting human influence on Climate Change, there is really nothing we can currently do to prevent a hurricane; its precursors, as with a pandemic (and, unlike wildfires), are beyond our capacity to control. We can only prepare for and react to it.

As seen in Fig. 6.1, we have plenty of experience with hurricanes. Our current abilities regarding satellite tracking, air-reconnaissance, and computer- modeling have dramatically improved our predictive capacity to forecast time-of- landfall, wind speed, ocean-surge, and flood-extent. Of course, on-the-ground surveillance is also vital. Such forecasts and observations are invaluable for planning and executing effective emergency response. Nonetheless, Nature and human miscalculations (as well as human folly) often have their way of thwarting our best efforts.

P. A. Erickson, *Effective Environmental Emergency Responses*,
https://doi.org/10.1007/978-3-031-05893-6_6

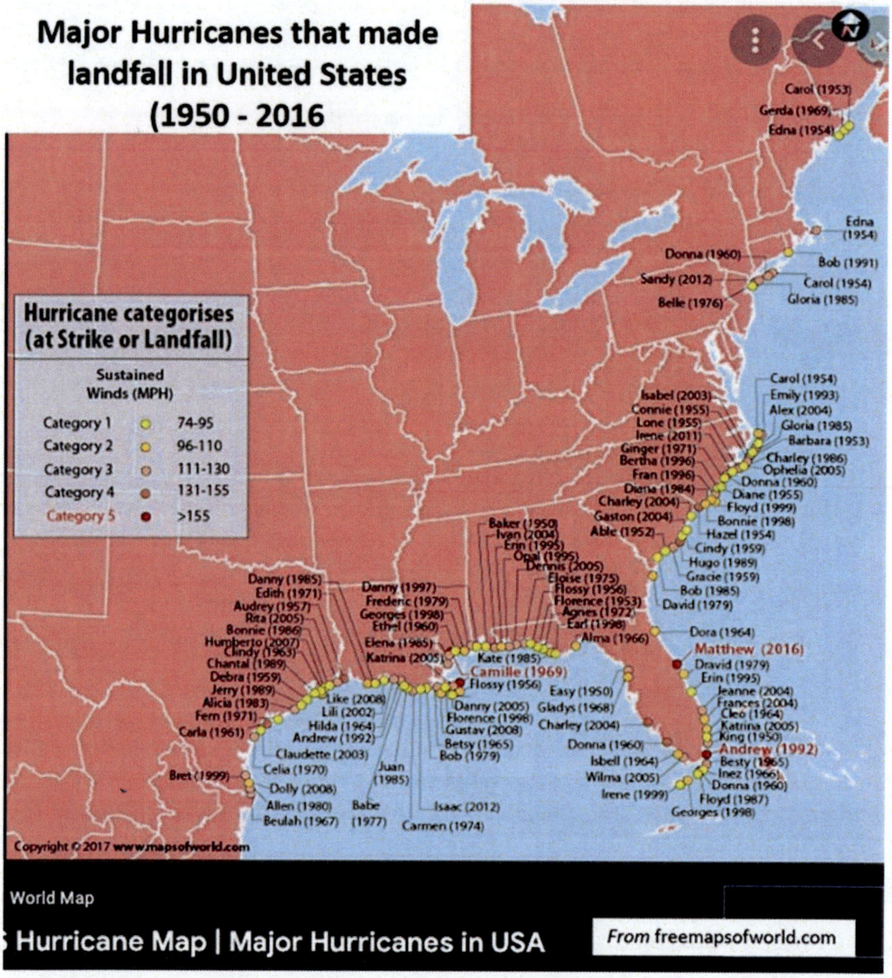

Fig. 6.1 American experience with hurricanes

For example, a Select Bipartisan Committee that investigated the Federal Response to Hurricane Katrina (August 2005) reported the following failures (U.S. Government Printing Office, February 15, 2006a, b):

- A perplexing inability to learn from Hurricane Pam (an exercise in preparedness, 2004) and other exercises
- Levees not built to withstand the most severe hurricanes
- An incomplete evacuation that led to deaths and tremendous suffering
- A complete breakdown in communications that paralyzed command and control and made situational awareness murky at best

- The failure of state and local officials to maintain law and order
- Haphazard and incomplete emergency shelter and housing plans
- An overwhelmed FEMA logistics and contracting system that could not support the effective provision of urgently needed supplies

The 3rd largest hurricane in U.S. history, Katrina destroyed over 800,000 housing units; 1.2 million were evacuated, and over 1800 died.

A study of those (over 200,000) who were evacuated from New Orleans to Houston revealed some telling characteristics:

- 55% did not have a car or a way to evacuate
- 68% had neither money in the bank nor a useable credit card
- 57% had total household incomes of less than $20,000 in the prior year
- 76% had children under 18 with them in the shelter
- 77% had a high-school education or less
- 93% were black
- 67% were employed full- or part-time before the hurricane (Heldman, 2011)

I do not include these above admitted failures necessarily as pejoratives; as I have emphasized in earlier chapters, mistakes are excellent learning-tools (if they are not repeated). The inequities, however, are clearly needed reminders that even heroic emergency-response efforts are necessarily reflective of both historic and persistent, insidious social-prejudices and attitudes … prejudices and attitudes, I emphasize, that are not regionally confined but can be found helter-skelter in America.

The percent of American households without registered cars (rather handy for evacuating) decreased in almost linear fashion in the period from 1980–2017, with (understandably) higher percentages in densely populated cities and lower percentages in cities with low populations. However, in that same period, another pattern emerges … white households not having cars ranged form 8% to 6% over that 37-year period; black households not having cars ranged from 24% to 14%, a rough proportion of about 1white per 3 black to 1 white per 2.3 black (National Equity Atlas, 2021). What's going on? Maybe not easy for blacks to receive loans? Or maybe blacks are typically at the low end of income?

My emphasis on a holistic approach to emergency planning and response is simply to expand our practical understanding of the actual physical- and social-scope of any emergency and, thereby, not overlook or under- estimate the relevance of interconnected causal-chains, including those activated not by Nature alone, but also by human behavior. Katrina exacerbated preexisting, debilitating inequities; it did not invent them.

Figure 6.2 is an overview of certain causal-chains that apply to hurricanes in general (whether 'large or small', historic or not, and coastal or inland). Because I have already commented briefly on some of its inequities (which I will revisit), I begin with evacuation, typically among the earliest first-responses to any hurricane.

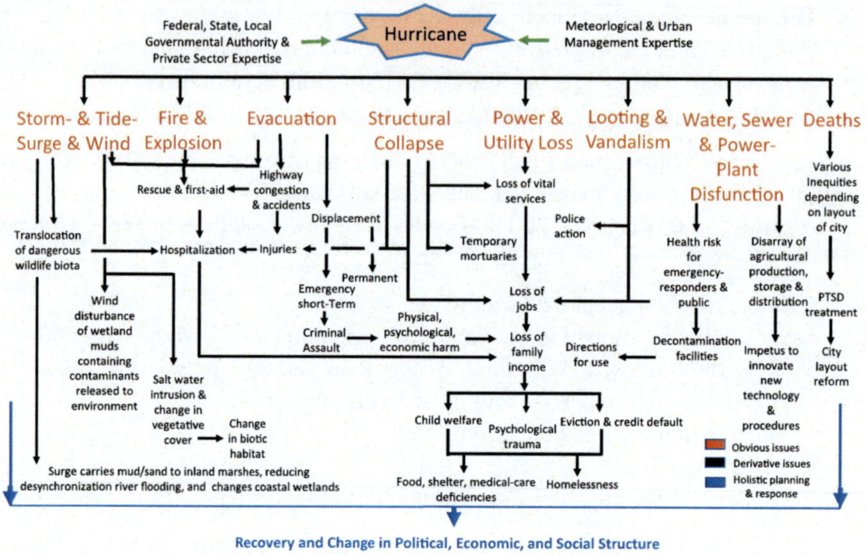

Fig. 6.2 Interconnections of causal chains associated with hurricanes

6.1 Evacuation

Much attention was given over past decades to the many factors that influence the success of evacuations, especially regarding the expected psychological reactions of drivers suddenly faced with the various constraints imposed by clogged highways, including panic, disorientation, fear-induced confusion, and anti-social behavior. After all, we have all experienced the misuse of break-down lanes by some frustrated, and most likely, angry drivers. However, early (and continuing) studies have indicated that such expectations are not founded on facts … that hurricane-evacuation is not panic-driven … that the more significant problem is to get people to evacuate in the first place (Auf der Heide, 2004).

Nonetheless, I note in Fig. 6.2 under 'highway congestion & accidents' simply because such things do happen, even if the probability may not be greater than in ordinary heavy traffic. I should also add, as a practical axiom, that it only takes one reckless driver (whatever the distraction or motivation) to cause a highway catastrophe under any circumstance.

Additionally, in many hurricanes, side routes to designated main highway-exodus express-ways are often along coastal roads subject to sudden storm-surge floods (Fig. 6.3) that can overwhelm even the most cautious driver. But those same storm-surges can also transport large amounts of sand and mud into inland wetland areas that provide numerous, significant environmental functions, including flood-desynchronization (Fig. 6.4), which involves their bleeding-off excess riverine flows and thereby reducing downstream floods. However, if those wetlands are

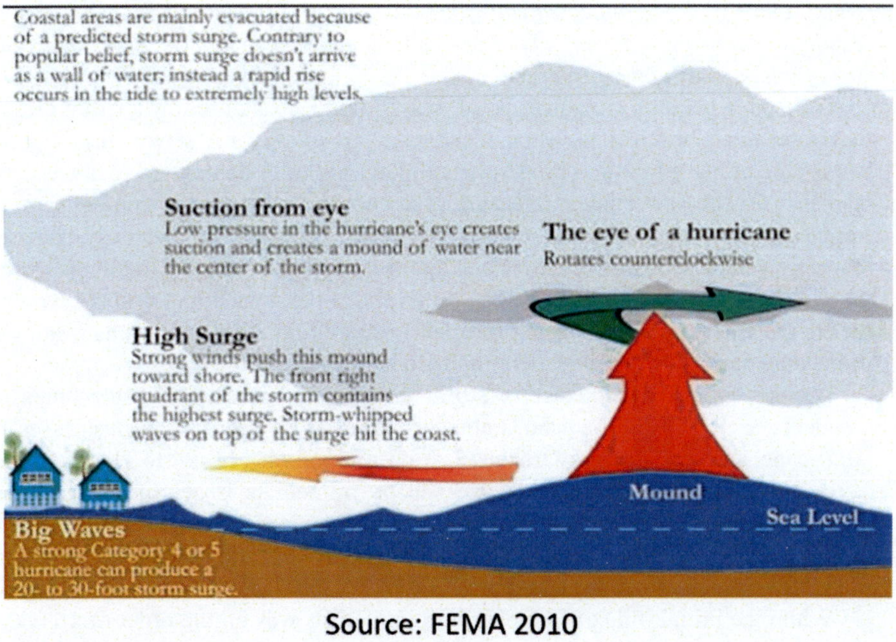

Fig. 6.3 Storm Surge

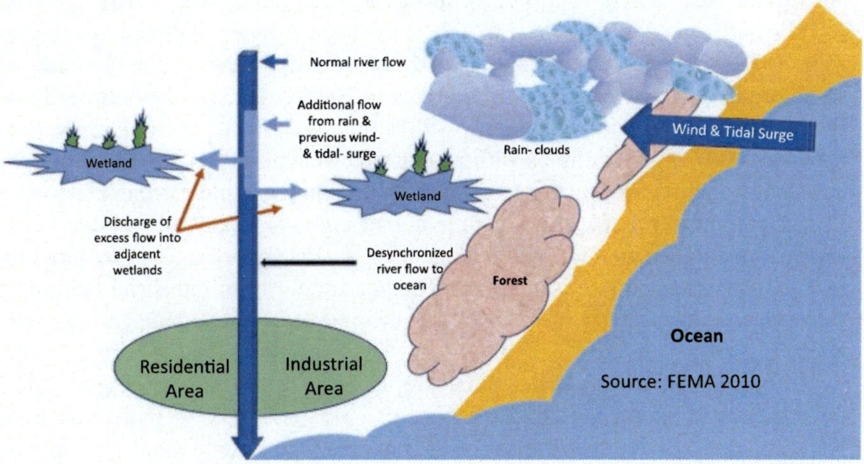

Fig. 6.4 Wetland desynchronization of river-flood

filled with surge-transported sand and mud, they cannot prevent subsequent riverine floods. This is an excellent example of how one hurricane-related impact (storm-surge) has a negative impact on local evacuations, but also becomes a cause of subsequent riverine floods … *or, need I say (more precisely),* just one example of what I mean by *interconnected causal chains …* and (I presume you insist) also an example of *how even one impact of one current emergency may become the precursor of a later, subsequent emergency.*

Perhaps. I belabor the above point, but so be it: coastal roadways that closely navigate along beachfronts often serve as rights-of-way for sewer lines. Storm-surge (including wind- and tidal-surges) inflect not only damage to roadways, but can also tear-up sewer lines, releasing sewer contaminants to the immediate environment. These contaminants, mixed with sand, water and mud already contaminated by vehicular exhausts are carried by the surge inland and are finally deposited in inland wetlands and rivers, thereby presenting a possible (if not probable) public health hazard. Again, a possible precursor to some future emergency, as well as an immediate hazard for those caught in the local hurricane-trajectory.

Another major consequence of storm-surge is, of course, the displacement of residents, especially those who, being displaced from homes and businesses by Hurricane Katrina (in New Orleans, La) on a temporary basis (to Houston, Tx), were provided emergency shelter and succor. Immediate issues of concern were high temperature and humidity in the Houston astrodome, as well as limited food and potable water, absence of basic sanitation facilities, and sewage-contaminated floodwaters (Editors, 2005).

While the initial influx of La evacuees into Texas was on the order of 16,000, a maximum of about 250,000 was finally achieved; about 150,000 of those took up permanent residence which quickly spawned a public perception of rampant criminal-behavior among the La evacuees, including rape and murder. However, a subsequent study did not support for this contention (Hamilton, 2010).

It is more than a little curious that the La evacuees were so quickly characterized as infested with criminals. A possible explanation, of course, is the fact that, while some predominantly white neighborhoods in New Orleans were entirely flooded, almost all neighborhoods having 75–100% black residents were entirely flooded (Logan, 2008). "Even in times without extreme weather disasters, the United States has an abysmal record Racism is a common breeding ground for misinformation.

It is certainly true that high-income homes are ravaged by natural disasters, but low-income homes are more often devastated because poor families are more likely to be concentrated in flood-prone areas and/or are closer to industrial facilities that can release hazardous chemicals due to severe storm-damage (Krause & Reeves, 2017).

Unfortunately, in the U.S., racial prejudice is often the underlying cause of poverty which, in turn, subjects people of color not only to the devastation of Nature's periodic furies but also of humans' persistent industrial pollution (Maxwell, 2018).

Despite the stubborn persistence of racism, The Strength in Diversity Act of 2020 (H.R. 2639)[1] has had a major impact on the increase in American diversity.

[1] Strength in Diversity Act of 2020 (H.R. 2639) provides funding and expertise to support school districts that are voluntarily developing, implementing, or expanding school diversity initiatives, and changing demographics in birth- and immigration-rates.

The numerical superiority of white populations has declined steadily over the past 40 years, from a total of over 11 million in 1970–1980 to less than 3 million in 2000–2019 (Frey, 2020).

It is interesting to note that, in this same period, American diversity has generally increased, though with regional differences: in some cities, diversity decreased (Fig. 6.5).

Despite regional differences, the fact remains that almost 70% of the largest American cities are more racially and ethnically diverse than they were a dozen years ago (McPhillips, 2020).

From an equity perspective, this would appear to be good news. However, the fact remains that many city-amenities (e.g., parklands, recreational facilities, municipal services, etc.) remain inequitably distributed. The issue is not to rebuild based on old prejudices and practices inherited from previous generations, but rather, to *revitalize* our cities with the new technologies and sensitivities of the twenty-first century.

Revitalization is the opposite of gentrification: the former consists of the reinforcement of social networks, neighborhood services and local businesses, and the maintenance of the existing neighborhood's character; the latter (which is too commonly a characteristic of contemporary America) is the refurbishment of the existing neighborhood in such a way as to cause the displacement of the poor by the privileged (Harries, 2017).

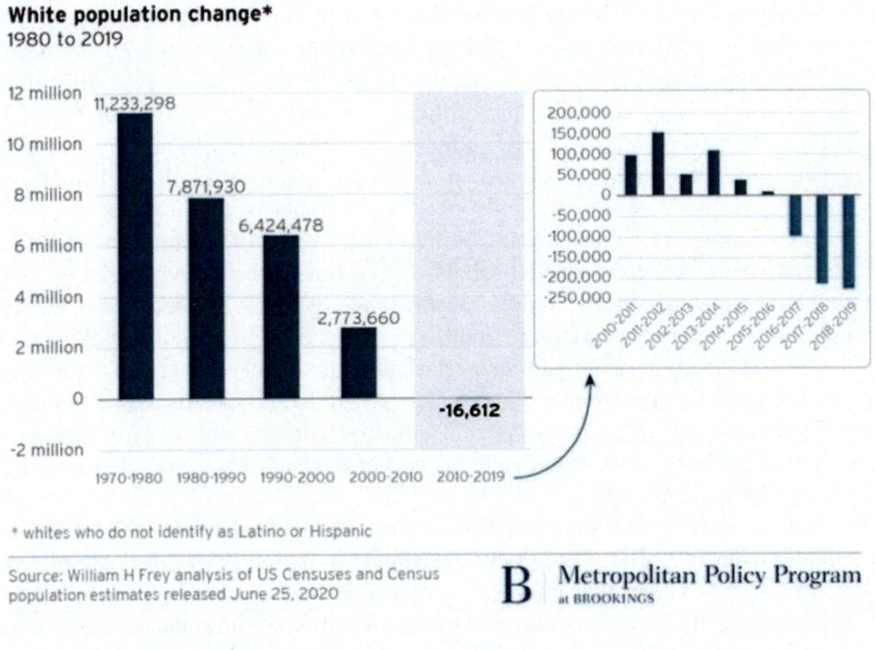

Fig. 6.5 American diversity

6.2 Disarray of Food and Other Production

Hurricane Katrina's devastation of agribusiness included the following:

- \$882 million losses in total crop, livestock, and aquaculture production in Southeastern regions.
- In Mississippi, where facilities were either damaged or without power for an extended period, an estimated 6 million chickens were killed, and 2400 poultry-barns were damaged; another 200,000 chickens were killed in Alabama. Estimated broiler losses estimated at \$15 million.
- USDA estimated timber damage on over 5 million acres in Mississippi, Alabama, and Louisiana. Note: dead and damaged trees can become hazardous fuels for wildfires as a haven for forest insects and diseases.
- Throughout the region, export traffic congestion as well as food-storage facilities became major problems requiring USDA ameliorations via loans and price-support measures. (Congressional Research Service, 2005)

Please note that the forest debris caused by high-wind damage was assessed as a possible precursor to a future wildfire … yet another example of *a consequence of one emergency becoming the potential cause of a future emergency.*

Also note that I have linked 'impetus to innovate new technologies & procedures' to 'disarray of food & other production'. I will pay much more attention to this linkage in Chapter? (Climate Change), but I think it useful at this point to introduce a topic that has world-wide significance.

Consider the fact that agriculture was invented about 10,000 years ago. In the following few centuries, humans steadily employed a variety of new technologies to improve agricultural efficiency, including the use of tamed animals to till soils, the use of natural and, subsequently, artificial fertilizers, the rotation of crops, and hybridization of plants. Over the following few generations, we added insecticides, genetic modification, hormonal growth-accelerants, and antibiotic medicines. Most recently, we have invented wholly new technologies whose names have yet to become commonly vernacular, including indoor vertical-farming, robotic harvesters, automatic waterers, seeding robots, digital sensor technology, blockchain and supply-chain technologies, remote sensors, and UAV_s (Unmanned Aerial Vehicles) used for the Precision Agriculture industry (applications of artificial intelligence) in the United States and internationally. Basically, the history of agriculture is largely dominated by the need to increase productivity. Today (2021), the world population is 7.9 billion; by 2025, it will be 8.2 billion. Increased productivity remains an urgent need, but *the transport and storage of agricultural production is of ever-increasing importance.*

Indoor vertical farming was simply an idea proposed in 1999 by Dickson Despommier and his students who envisioned a skyscraper-farm that could feed 50,000 people. Today, that idea has become a practical technology for world-wide application in urban centers that obviates the devotion of thousands of acres to agriculture as well as the myriad problems associated with the transport of fresh food from distant, traditional farmland to hungry city-dwellers. This proven technology, which has been developed over the past 20 years, when coupled with even more recent development of biological sciences (e.g., cloning, stem-cell technology, and genetic engineering) is yet again revolutionizing not only the growing of plants, but also of animal tissue.

Over the past several years, we have begun to grow animal meats in the laboratory rather than obtaining meat by slaughtering animals. Small amounts of tissues are painlessly taken from live animals and subsequently cultured to produce fibers that bulk up to form that muscle tissue we call meat.

Stop and think. These very recent developments in the biological and material sciences lead to new technologies that, in turn, are used to deal with day-to-day problems. Just as improvements in air quality as the result of the Covid 19 pandemic (Chap. 5) serve as an impetus to the development of electrically powered automobiles, so does the set of problems related to Katrina's disarray of food production, storage and distribution serve as an impetus to the on-going development of vertical-agriculture and even meat-production, with both serving, in different ways (which I have yet to talk about), with the phenomenon of climate-change (Chap. 9).

Yes, I know that, as an old teacher, I tend to repeat myself … so be it. But if we remain content in our thinking old ideas simply because they have become (with age) unquestioned truths, we will have lost that basic human-urge to learn anything new and, thusly, to learn how to deal effectively with a world vibrantly alive with constant change and challenge. Science and its offspring, technology, must replace what is only unexamined custom.

6.3 Displacement and Its Consequences

Whether hurricane-induced evacuation results in temporary or permanent displacement, major attention must be made to psychological stress of evacuees, especially regarding the time required to implement effective treatment.

Victims of several hurricanes that occurred over a seven-week period in Florida (2004) were evaluated (Acierno et al., 2011) 6–9 months after their evacuation-experience. Twenty percent has suffered relatively minor property damage; 8% had suffered catastrophic property damage. Fifty percent of those who had been employed prior to the hurricanes had lost time at work and/or had lost their jobs. Of those with prior medical problems, 14% had difficulties getting their medicines, and 9% experienced serious difficulties obtaining essential medical equipment. Even after 6–9 months, many reported a variety of after-effects, including anxiety, nervousness, worriedness, loss of appetite, sadness, sleeping difficulties, and reduced capacity to work or study.

These above and other symptoms are variously included in the following categories of psychological stress: posttraumatic stress disorder (PTSD), Generalized Anxiety Disorder (GAD), and Major Depressive Disorder (MDD). Untreated, such conditions can seriously cause personal and societal dysfunction (Gray, 2019).

So, what's the cost of treatment? It's highly variable, depending on the specific diagnosis, the different medicines required, the time taken for counseling and medical treatment, whether the patient has veteran-status or other insurance. Cost per year can easily range from roughly \$1000 to \$5000 or more. Psychotherapy sessions required for depression may extend from several weeks, or months, or even years and can cost \$100 per hour.

It should not be surprising that, whether marginalized by poverty, color, or any other social or institutionalized prejudice, many Americans are twice as likely to experience psychological distress than those who live with incomes over twice the poverty level (MHA, 2021), which is, in 2021, \$26,500 for a 4-person household. Of course, you don't have to be marginalized to experience economic disaster because of psychological trauma: consider those we rightly identify as heroes, the many 'first responders' and 'essential personnel' we daily rely upon.

Median incomes[2] vary greatly from state to state, over time, ethnicity and/or race. Such variances result in inequities in emergency-response efforts, including those dependent on on-site personnel and equipment, as well in differences of impacts on victims, especially regarding the typically limited duration of response-related treatment and resolution of psychological trauma.

It is therefore important to emphasize that trauma, which is a common consequence of many types of emergencies, may often have long-term effects that (as with hurricanes and pandemics), if ignored or otherwise unresolved, can become precursors to subsequent emergencies, including significant personal and anti-social behavior and even violence. After all, among our most celebrated scientists, it is a deeply held conviction that the two most baffling phenomena to human understanding is the Universe itself and how the Human Mind works. We have come a long way so far, but we have a very, very long way yet to go.

6.4 Overview of Fig. 6.2: Holistic Response

As with the Covid-19 pandemic, we obviously can only react to (rather than prevent) hurricanes. However, it is appropriate to consider what reactions can be vastly improved.

- Should evacuations be made mandatory at the regional level and not by local authorities?
- How best can we ensure the universal availability of transport for evacuees?
- What steps should be taken to ensure that temporary shelters are adequate regarding availability of food, water, medicine, safety, and sanitary facilities.?
- How do we revamp the layout of cities to facilitate effective and equitable evacuations?
- How do we ensure both short- and long-term psychological counseling?

[2] See Appendix 2 for additional information on income variance in the U.S.

- What mitigation measures might be taken to return silted, inland-riverine-wetlands to their normal environmental functions?
- How might State and Local authorities facilitate the adoption of 'vertical farming and husbandry'?

Suggested Exercises for This Chapter

1. Imagine that you and your partner, along with 2 children, have received an order to evacuate 24 h before the landfall of a hurricane. Also imagine that, having no car, you have received telephoned advice to arrange transport with neighbors. Write an essay (and/or give an oral report) that describes your reaction to that advice.
2. Using the internet, compare the response to Katrina (Aug. 23–31, 2005) to the response to a subsequent (2020–2022) hurricane in Louisiana. Give particular attention to improvements and (if any) persistent failures.
3. Choose a partner for a classroom debate (or a joint paper) regarding the pros and cons regarding which governmental authority should have the sole power to initiate evacuation.
4. Write an essay on how you would manage zoning for domestic residences with respect to properties prone to storm-surge destruction.

References

Heldman, C. (2011, August 29). *Hurricane Katrina and the demographics of death.* Sociological Images.

MHA. (2021). Mental Health America, Inc. Black and African American Communities and Mental Health.

National Equity Atlas. (2021). *Percent of households without vehicle, USA, 1980–2117.*

U.S. Government Printing Office. (2006a). *Federal response to Hurricane Katrina: Lessons learned.*

U.S. Government Printing Office. (2006b). *A failure of initiative: Final report of the select Bipartisan Committee to investigate the preparation for and response to Hurricane Katrina.*

Chapter 7
Domestic Terrorism

The 107th Congress of the U.S. expanded the definition of terrorism to cover "domestic terrorism" (Section 802 of the USA PATRIOT Act [Pub. L. No. 107-52]) as follows:

> *A person engages in domestic terrorism if they do an act "dangerous to human life "that is a violation of the criminal laws of a state or the United States, if the act appears to be intended to: (i) intimidate or coerce* a civilian population; (ii) influence the policy of a government by *intimidation or coercion; or (iii) to affect the conduct of a government by mass destruction, assassination or kidnapping. Additionally, the acts have to occur primarily within the territorial jurisdiction of the United States and if they do not, may be regarded as international terrorism.* (U.S. Congress, 2001).

While domestic terrorism is a modern term, it has a long American history that includes, among others, Shay's Rebellion (1786–1787), New York City Draft Riots (1863), Richmond Bread Riots (1863), and the Wilmington Insurrection (1898), the assassination of 4 Presidents [Abraham Lincoln (1865), James Garfield (1881), William McKinley (1901), and John Kennedy (1963)] as well as the attempted assassinations of Theodore Roosevelt (1912) and Ronald Reagan (1981).

I purposely leave out of this sordid list of American violence the Civil War … which, as a war, involved hundreds of thousands of combatants and not the paltry numbers of citizens who, intent on inflicting harm on fellow citizens for whatever diverse and nefarious reasons, hardly constitute an army.

Domestic terrorists range from single individuals to small groups. However, whether as an individual or as a cabal, they can cause great harm and thereby elicit the growing concern in America that they potentially present a national risk that cannot be (and, historically, has not been) ignored.

As evidenced by the 2021 U.S. Capitol Insurrection, history does seem to repeat itself. But not exactly.

On the one hand, a good number of those we saw on TV that day do not necessarily qualify as terrorists but, rather, as momentary exhibitionists caught up in that crowd-frenzy typically witnessed at rock-concerts and other spectaculars.

P. A. Erickson, *Effective Environmental Emergency Responses*,
https://doi.org/10.1007/978-3-031-05893-6_7

On the other hand, technology has evolved thunderously over the past several centuries; even the most primordial, devilish proclivities of humans can be fast-tracked by weapons that, previously, could not be even imagined (e.g., rapid-fire assault rifles; instant, world-wide communication, easily available manuals on explosives, bio- and chemo-toxins, remote tracking, etc.). Thus, the wisdom of today's Federal authorities in paying attention (as those in the past always have) to domestic terrorism as a potential National emergency.

Let me underscore a simple fact about America: whatever the social or political issue under discussion, there is (and always has been) extensive disagreement, sometimes very mild, sometimes vituperative, and often very divisive. It is not disagreement that defines a terrorist. Disagreement is as American as "apple pie". A domestic terrorist is whoever *purposely translates disagreement into actions that directly harm fellow citizens*.

Because the words 'liberal' and 'conservative' are so dominant in American social-life, I hereby suggest (surely to the chagrin of professional lexicographers) two non-political definitions that go beyond partisan politics and are, so I think, more useful: Conservative (adj.): a way of thinking that is strongly influenced by traditional ideas, values, and behavior; Liberal (adj.): a way of thinking that is strongly influenced by contemporary ideas, values and behavior. In previous chapters, I have discussed the value of both old and new ways of thinking … that both old and new ways of thinking are treasure-houses to be mined for their practical, contemporary relevance.

Considering these two (perhaps too simplified) definitions, I think you can easily understand that a person may be conservative regarding one issue and, at the same time, be liberal on another, or for that matter, change from one to the other on the same issue over time. People are not robots born with a single, unchangeable program … though, of course, experience may sometimes contrive a way to do just that.

Figure 7.1 incorporates my above definitions. It is a variant bell-curve depicting to the left 'liberal dissatisfaction' with a variety of issues (*in green figure*) and, to the right, 'conservative dissatisfaction' with the same issues. Dissatisfactions to either side of Fig. 7.1 range from no dissatisfaction to extreme dissatisfaction. It is the latter category (red) that represents the small (but extremely dangerous) populations of domestic terrorists among both conservative and liberal factions.

What motivates such extreme behavior? There are numerous, possible explanations, including financial instability, mental health issues, personal traumatic experience, recruitment by mentors, propaganda, social bonding with other extremists, childhood experience, family dynamics, and social backgrounds involving societal marginalization (Brown et al., 2021).

Certainly, anger has often been seen as a precursor to aggressive behavior; there is research-evidence that shows that "about 90 percent of aggressive incidents are preceded by anger. However only 10% of anger-experiences are actually followed by aggression. People often *want* to act aggressively when angry but, fortunately, most do not *actually* take aggressive actions" (Kassinove, 2012). Regarding any individual, who really knows? I leave that question to

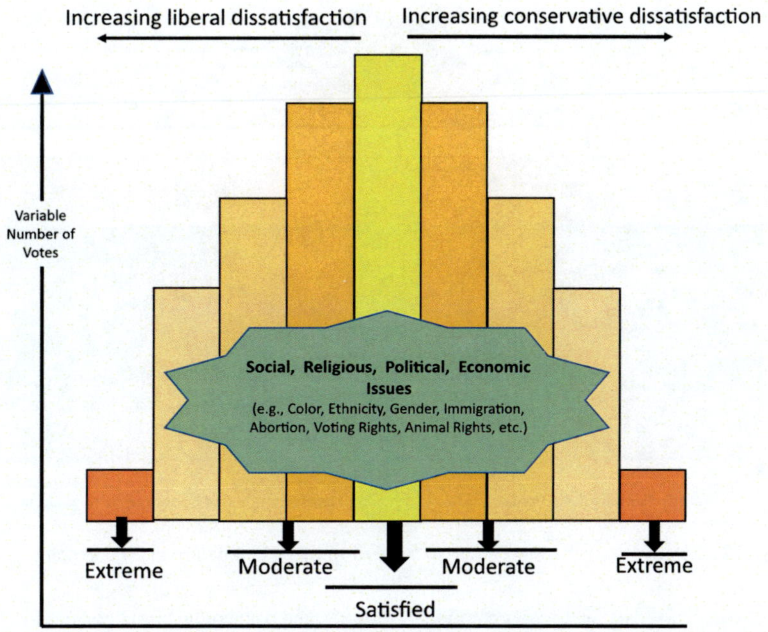

Fig. 7.1 Liberal and conservative dissatisfactions

professionals in the psychological, social, behavioral, and neurological sciences and focus, rather, on means and consequences of 'home-grown', societal terror. Before I do that, however, I want to underscore an important fact about the emergencies I have thus far discussed in the previous 3 chapters: specifically, that any emergency may be conceptualized as a situation ultimately defined by the complex *interactions between human behavior and Nature's physical, chemical, and biological phenomena*.

As I discussed in Chap. 4, wildfires are most often the result of human carelessness; those started by lightening are relatively few. Likewise, regarding pandemics, Nature obviously creates the pathogen, but it is humanity that spreads it so quickly throughout the globe, much more efficiently than wind. Nature produces the hurricane, but humans determine where they live and where and how thy build their roads and domiciles. Nature doesn't contrive Incidents of domestic terror; they are the sole results of human behavior. As shown in Fig. 7.2, human behaviors often play a significant, if not the leading role as precursors to subsequent emergencies. Unfortunately, humans also suffer greatly from those same emergencies, including those we describe as emergency first responders.

Emergency response services may become involved in terrorists acts in two ways: first (and most obvious), simply because they respond to any community emergency, whatever its cause; second (and too often ignored), because they themselves may become the object of terrorists acts or otherwise unwittingly become pawns in terrorist strategy. To emphasize the difference in these two types of involvement, it is instructive to consider that, while 'intimidation, coercion, or ransom' may in fact be key motivational dimensions of docu-

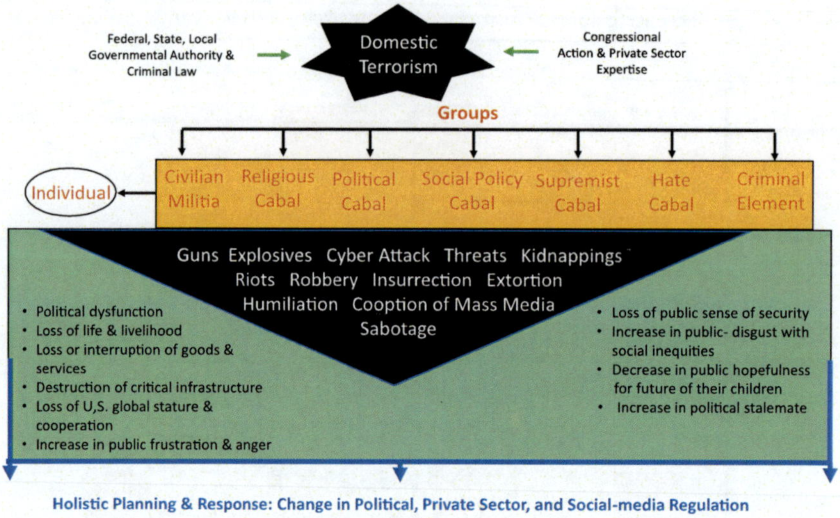

Fig. 7.2 Relative contributory causative roles of humans and nature in emergencies

> mented terrorists' acts, other emotions, volitions, and psychological states can as well serve to unleash wanton disregard for human life – including revenge, anger, and frustration and even (albeit perversely and pathologically misguided) a sense of excitement or challenge. (Erickson, 1999).

Response services become objects of terrorist violence for several and diverse reasons:

- Because they are high-profile targets of opportunity
- Because their dependable response to an emergency makes them predictable and therefore susceptible to the advanced planning of those intent on violence
- Because they are so central to community safety and health, and yet finite in number they become that first shield of defense to break apart before releasing a more concentrated onslaught on the community proper
- Because their field operations necessarily entail physical disruptions of normal traffic flows and so occupy the attention of other community services and the public that they can be used both as tactical bottlenecks and feints in a many-layered stratagem for city-wide chaos. (Erickson, 2006)

As these examples illustrate, to focus on intimidation, coercion, ransom or any other objective as the sole or even primary motivation of domestic terrorism is, perhaps, to define the risk presented by terrorism to emergency services (and, thence, to the public at large) too narrowly, essentially ignoring not only the range of human motivations for perpetrating horror, but also a basic corollary of any humanly contrived and directed violence – that if you play a tactical role against the interests of an adversary, then you yourself become of special interest in that adversary's strategy violence – a corollary illustrated by a study of domestic terror, in the period January -August 2020, showing 18% of acts of terrorism committed by far-right extremists targeted first responders (i.e., governmental, military, and police); 58% of incidents committed by far-left extremists also targeted first responders (Jones et al., 2020).[1] In short, when it comes to domestic terrorism, first responders are as likely to become first-targets as anyone else.

7.1 Range of Weaponry

Regardless of underlying motivation, a terrorist act is most brutally characterized by the willful use of weapons of mass and indiscriminate murder and destruction. Historically, the typical terrorist weapon has been an explosive device. Today, there is a wide range of weaponry that is increasingly available to terrorists. The U.S. Department of Justice and the Federal Emergency Management Agency have used the acronym B-NICE to denote this range, the letters standing for Biological, Nuclear, Incendiary, Chemical and Explosive agents.

Biological agents include microorganisms (i.e., bacteria, rickettsia, and viruses) and byproducts of microbiota (e.g., Botulinum toxin) that can be disseminated within human populations as atmospherically released aerosols and particulates, the contamination of food, water, and other common-use items (e.g., cosmetics, clothing, and the use of living carriers (humans and animals).

Nuclear agents are of two basic types: actual thermonuclear devices and conventional explosive devices that structurally incorporate nuclear materials (i.e., 'radiological dispersal device', or RDD) and which could be used to disperse hazardous nuclear materials over an extended area. It is also possible to achieve the same effects of an RDD by detonating conventional explosives in the immediate vicinity of normal sources of nuclear materials (e.g., nuclear power plants or transport vehicles carrying nuclear cargo.

Incendiary agents include any mechanical chemic, and/or electrical devices specifically designed to start a fire. Whatever the arrangement of components, any incendiary device consists of an igniter or fuse, a container, and a flammable or combustible accelerator that, once ignited, serves as a source of fire for surrounding combustible materials.

Chemical agents may be classified in several ways. Categories commonly used by the U.S. Department of Justice and the Federal Emergency Management Agency include Nerve, Blister, Blood, Choking, and Irritating agents.

Given the Internet, any person with a computer can learn to use available technology to concoct a murderous weapon in their own home. And when I might not be able to find what I want, I can use instant communication-technologies to find somebody in the world who can tell me where to go to get it. Of course, when it comes to guns, that's a convoluted story, having been so for many past decades and into contemporary times, a political stew.

[1] See Appendix 3 for additional information on terroristic incidents in U.S.

7.2 Psychological Impacts on Victims and Society

> The psychological and emotional aftereffects of a major disaster are more severe and longer lasting when the disaster results in significant numbers of fatalities, seriously injured victims and destroyed businesses and homes. Those most personally touched are likely to experience the greatest suffering. Surviving victims and bereaved families will experience a range of short- and long-term impacts that are emotional, physical, financial, and legal. The characteristics of a disaster, which include a lack of warning, extreme threat to life, exposure to trauma, and uncontrollability, also contribute to the *severity and duration of the victims' psychological reactions*. (U.S. Department of Justice and American Red Cross) [Italic emphasis by author].

As discussed in previous chapters, the 'severity and duration' of trauma are central considerations in any emergency because psychological impacts are typically long-lasting and require extensive treatment. Without such treatment, victims may themselves develop severe anti-social attitudes and act accordingly by promoting such attitudes in others and/or by, themselves, committing violence on others. This is particularly true regarding young children who have been abused or otherwise subjected to violence.

> Child abuse and neglect and other ACEs (adverse Childhood Experiences) can also have a tremendous impact on lifelong health opportunity and wellbeing if left untreated. For example, exposure to violence in childhood increases the risks of injury, *future violence victimization and perpetration*, substance abuse, sexually transmitted infections, delayed brain development, lower education attainment, and limited employment opportunities. (CDC, 2021 [Italic emphasis by author]).

A study of children and violence by the DOJ (U.S. Department of Justice) found the following:

- 60% of American children were exposed to violence, crime, or abuse in their homes, schools, and communities.
- Almost 40% were direct victims of two or more violent acts, and one in ten were victims of violence five or more times.
- Almost one in 10 saw one family member assault another family member, and more than 25% had been exposed to family violence.

The study concluded that "Whether as victims or witnesses, children's exposure to violence is often associated with long-term physical psychological, and emotional harm. Children exposed to violence are also at a higher risk of engaging in criminal behavior later in life and becoming part of a cycle of violence" (Finkelhor et al., 2009).

Of course, adults are also variably susceptible to radicalization depending upon numerous experiential factors, including lack of socio-economic opportunities, marginalization, poor governance, violations of human rights and the rule of law, prolonged and unresolved conflicts, imprisonment where radicals are held, witnessing torture or brutal killing of a relative or friend, experiencing severe humiliation, or being denied rights over a sustained period (Milani, 2017).

That domestic terrorists might evolve from the previous experience of violence and from the psychological consequences of social inequities would appear to be at least (let us say) marginally reasonable. Assuming the reasonableness of that idea, consider also that it is unquestionable that the human species is a favorite contender for the title of 'Most Physically and Psychologically Complex Species' on earth. The conclusion must be that domestic terrorism is not going to disappear. Among us, there are saints as well as monsters, …. and, of course, all the rest of us who must decide if it's safe to go outside.

Recently, our basic approach to dealing with domestic terrorism has been called the "hard security approach" which is the use of force, including surveillance, policing, and the implementation of antiterror laws … basically, the focus is on the abrupt prevention of terrorism. This approach contrasts with the more recent "soft security approach" which focuses on those factors that enable extremist-ideas to grow and be sustained (Subedi, 2017). I suggest that both approaches are needed.

7.3 Social Consequences

Domestic terrorism has been and will continue to be a factor in American daily life simply because humans are humans. However, the never-ending progress of time and technology always brings novelties, including inspiration by historical and newly evolved ideologies, radicalization via the internet, and easy access to weapons (Bergin et al., 2019).

A review of data on terrorist attacks in the U.S. Between 1970 and 2013 shows a distribution of target-type (structures and persons) for several categories of terrorist attack as follows:

Abortion related: 94% (buildings), 6% (persons)
Educational related: 78% (buildings), 22% (persons)
Governmental related: 73% (buildings), 27% (persons) (Miller, 2014)

Given the loss of life, psychological horror, and financial losses associated with any terrorist attack, certain modern circumstances deserve serious consideration for their potential long-term and pervasive influence.

For example, the population of the United States in 1790 was 3.9 million; as I write this sentence, it is a little over 328 million. Back then, news traveled across the countryside at the speed of a horse-trot or boat; now it traverses the globe 7.5 times in 1 s. Back then, human knowledge was stored in printed books available only to the few; now it is digitized and available to anybody in a matter of seconds.

Sure, perpetrators of terrorism make use of the possibilities that contemporary technology presents in their planning, coordination, access to means, and execution of horror. But precisely the same informational resources are rightfully available to the general public.

The real power of domestic terrorists is not in their numbers but in their ability to gain the limelight for their (whatever) complaint. Once in the limelight, that complaint can demand attention (even if subliminally) and, possibly, serious consideration. Thus, a significant consequence of a terrorist attack is not just the murder of innocent people, but *the initial recruitment of individuals to the murderous cause itself.* I use the phrase "initial recruitment" here to emphasize that recruitment itself is a multi-step process that typically begins with a sudden awareness of unconventional thinking and behavior that sparks even a modicum of interest into progressively attentive inquiry of similarly attracted individuals.

Today, social networks established and maintained by both mass and social media are fundamental tools available to activists to instigate, recruit and/or coordinate other individuals or even loosely coordinated groups intent on personal or public violence (Busher, 2015).

The roles that mass and social media play regarding social violence are not going to disappear; in fact, they may be enhanced by likely advances in communicational technology. For example, while we have recently relied on the virtual communication provided by Zoom technology for home education and home-based work during the Covid pandemic, that technology may well be quickly supplanted by holographic technology (as in Star Wars) which will add a whole new dimension to digital communication via both mass and social communications.

Regardless of the technological means that may promote social violence, any individual may act of violence may occur 'just for the personal satisfaction of it' and not with any intention of impressing others. However, whether by a cabal or an individual, the mere mass-media deluge of news regarding any terrorist act has its affects as shown in Fig. 7.2, including:

- Loss of public sense of security
- Decrease in public hopefulness for future of their children
- Increase in public frustration & anger
- Increase in public disgust with social inequities

Alone and in combination, these consequences have direct influence on political decision-making, especially regarding political dysfunction.

Former speaker of the house Tip O'Neill once declared that "all politics is local", a statement that has been variously translated over the years. I am presumptuous enough to translate it thusly.

First: politicians have to be elected; Second: in order to be elected, they have to satisfy their voters; Third: they are not going to satisfy their voters if those voters are angry with them.

So, it comes down to this: Does the politician pander to the extant biases and privileges of voters, or does the politician attempt to pursuade them as to a more socially beneficial course? That's a very difficult question. While the latter is the hallmark of a leader and, the former, the measure of a narcissist, one cannot be a political leader if not elected. That's the quandary.... how to get elected and be also a leader. Well, pardon me if I suggest that 'political stalemate' may be symptomatic of a political structure having too many narcissists and too few leaders in charge.

7.4 Overview of Fig. 7.2: Holistic Response

Domestic terrorism has nothing to do with Nature; it is the product of human nature, and its consequences are directed against fellow humans, individually and/or collectively. What might we consider as ways to reduce domestic-terrorism's toll on the American quality-of-life?

The constitutional 'right to bare arms' was written at a time when 'arms' consisted of what we would today describe as primitive. How do we reconcile that 'right' with public possession of modern assault weapons?

- How do we balance the ownership and display of guns between municipal, State, and Federal authorities?
- What steps might be taken to reduce childhood experience of violence?
- The misuse of mass-media regarding misinformation and just plain lies has become of major concern. How might we square the regulation of mass-media with the constitutional rights of 'free speech' and 'freedom of the press'?
- How might we distinguish between the regulation of 'social media' and 'commercial media'?
- How might we encourage commercial advertising to discourage anti-social biases, including (but not limited to) those regarding 'race', ethnicity religion, sexuality, and gender.

Suggested Exercises for This Chapter

1. There are many kinds of American institutions that grant a variety of loans that necessary for domestic and commercial purposes, including central banks, retail and commercial banks, internet banks, credit unions, savings, and loans associations, investment banks, investment companies, brokerage firms, insurance companies, and mortgage companies. Select any appropriate local institution and discuss (or write an expository essay) regarding possible changes in the process of awarding loans that might be essential for eliminating current inequities regarding 'racial', ethnic, or gender related biases in the award of loans.
2. Discuss (or write an expository essay) on the positive and negative consequences of regulating the use of social and commercial mass–media to reduce the promulgation of misleading and/or purposefully untruthful information.
3. Discuss (or write an expository essay) on the pros and cons regarding the efforts of local– and State officials to contravene Federal mandates regarding public health and safety.
4. In the United States there are serious concerns regarding increasing both the facilitation and restriction of the casting of votes. What's your feeling about this issue and explain your reasoning?

References

Bergin, P., et al. (2019) What is the threat to the United States? Terrorism in America after 9/11. *New America.*

Brown, R., Helmus, T., et al. (2021). *What do former extremists and their families say about radicalization and deradicalization in America?* RAND Corporation.

Busher, J. (2015) *What part of social networks play in radicalization?* (Paperback ed.). University of California Press.

Congress. (2001). Section 802 of the USA PATRIOT Act [Pub. L. No. 107-52].

Erickson, P. A. (1999). *Emergency response planning for corporate and municipal managers.* Academic.

Erickson, P. A. (2006). *Emergency response planning for corporate and municipal managers* (2nd ed.). Elsevier.

Finkelhor, D., et al. (2009). *Children's exposure to violence: A comprehensive national survey.* DOJ, Office of Justice Programs, Office of Juvenile Justice and Delinquency Prevention (Updated 2017).

Jones, S., et al. (2020). *Domestic terrorism in the United States.* Center for Strategic and International Studies (CSIS).

Kassinove, H. (2012). *How to recognize and deal with anger.* American Psychological Association.

Milani, L. (2017). *The process of radicalization.* Futures Without Violence.

Miller, E. (2014). *Terrorist attacks in the U.S. between 1970 and 2013: Data from the Global Terrorism Database (GTD).* START (National Consortium for the Study of Terrorism and Responses to Terrorism).

Subedi, D. (2017). Early warning and response for preventing radicalization and violent extremism. *Peace Review, 29*(2), 135–143.

Chapter 8
Earthquake

Human societies have utilized many different strategies for surviving the hazards inherent in the dynamic physical, chemical, meteorological, and biological processes of earth. Whatever the challenge, it was the malleability of human behavior that determined their success or failure. Some dealt with environment as it was, did not attempt to make changes to it but, rather, altered their behavior to adapt to it, often moving seasonally or permanently into different environments. Others made relatively small changes such constructing their own shelters and destroying or regenerating various kinds of plants. Over thousands of years, humans have made progressively major modifications, until finally becoming global architects capable of transforming whatever Nature provides into whatever humans may decide is most comfortable for themselves (Fernandez-Armesto, 2001).

While we have for so long concentrated on separating ourselves from Nature, we have continued to rely on our capacity to adapt our behavior, not to the dynamics of Nature but to our technologies, and to the increasingly complex social dynamics of our dense populations. Of course, there are still many of us who live quite close to Nature without access to modern conveniences. But most of us depend upon a plethora of advanced technologies and, day-to-day, upon manifold and convoluted social interactions; most of us haven't the faintest idea regarding survival in what we, in the comfort of our gilded cages, refer to Nature as 'wilderness' or, even worse, 'the zoo or park'.

In previous chapters, I have focused on interconnected causal chains, some of which result in both short–and long–term consequences of specific types of emergencies, and some that serve as precursors to yet other emergencies that typically do not receive media–attention from an easily distracted populace. Nevertheless, with respect to both precursors and subsequent consequences, human behavior (inclusive of both our actions and inactions, as well as our thinking and our doing) is often a significant contributor to the inception and magnitude of whatever we decide to describe as an emergency.

P. A. Erickson, *Effective Environmental Emergency Responses*,
https://doi.org/10.1007/978-3-031-05893-6_8

Metaphorically, it is time to view ourselves (regardless of historical religious or philosophical teachings) as tenants of earth and not its owners…. indeed, temporary tenants who…, like any tenant, may either trash or enhance their temporary abode.

In Chap. 2, I *suggested that the ethos of NEPA (National Environmental Policy Act) should be considered* in achieving a comprehensive, effective emergency management process that includes consideration of (1) precursors to emergencies and (2) both immediate and long-term, equitable remediations …. that such an approach would obviously require serious revamping of governmental and private-sector interactions as well as significant societal consensus.[1]

I also suggested that, while NEPA typically applies to construction–projects (highways, pipelines, etc.), the ethos underlying NEPA is based on the premise that human behavior and the dynamics of the natural environment are, in fact, intrinsically interconnected…. a premise that is applicable not only to construction–projects, but also to emergency planning and the execution of effective, comprehensive, and equitable response efforts. This chapter and the following focus on two types of emergencies that are of world–wide importance and which also clearly exemplify key interactions between human behavior and global dynamic processes.

8.1 On Earthquakes

The National Earthquake Information Center (of the U.S. Geological Survey) estimates that there are about 20,000 earthquakes on earth per year (averaging about 55 per day). Around 9624 people died worldwide in 2015 because of earthquakes (PolicyAdvice, 2021).

In the United States, it is estimated that roughly 12,000–14,000 earthquakes occur each year at a cost of $6.1 billion annually in building stock losses (FEMA, 2017). The deadliest earthquake in the US can be traced back to 1906. This quake happened in San Francisco, California, and destroyed most brick buildings and wooden Victorian structures. Nearly 30,000 buildings were damaged and 3000 people were killed (Simpson et al., 1981).

Most earthquakes are caused by the collision of continental and oceanic tectonic plates (Fig. 8.1). Both types are solid rock, but continental plates are lighter, whereas oceanic plates are heavier. They can be imagined as two pieces of eggshells floating on the surface of a wet hard–boiled egg, and, when they collide and grind against each other, the heavier one is subducted under the lighter plate. Of course, the earth is not hard–boiled; its center is a metal–heavy, molten mass which allows both plates to meander and periodically collide.

Figure 8.2 illustrates that a possible result of collision is the formation of a volcano such as Mount Saint Helens (Washington State) that catastrophically erupted on May 18, 1980. However, subduction of an oceanic plate does not always result in

[1] Refer back to Fig. 2.6.

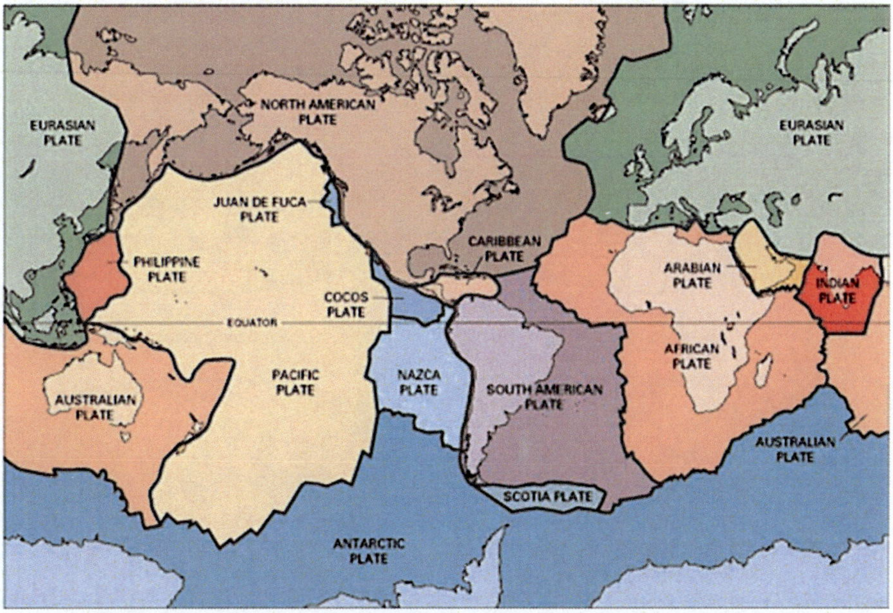

Fig. 8.1 Tectonic plates

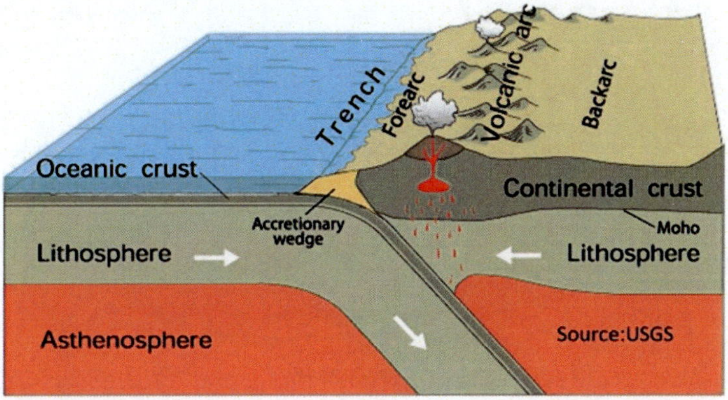

Fig. 8.2 Subdduction

the formation of a volcano. For example, the Himalayan Mountains are not volcanic in origin; some 40–50 million years ago two large landmasses, India, and Eurasia, collided. The pressure of grinding plates thrusts the collision zone skyward, thereby forming the Himalayan mountain-range. The same thing happens when you dig your toes into wet beach–sand and thereby create a pile of sand in front of your foot. Applied pressure has consequences!

Regarding the globe, the dynamics of tectonic plates cause continents to drift over geological time. As shown in Fig. 8.3, 250 million years ago, all the continents

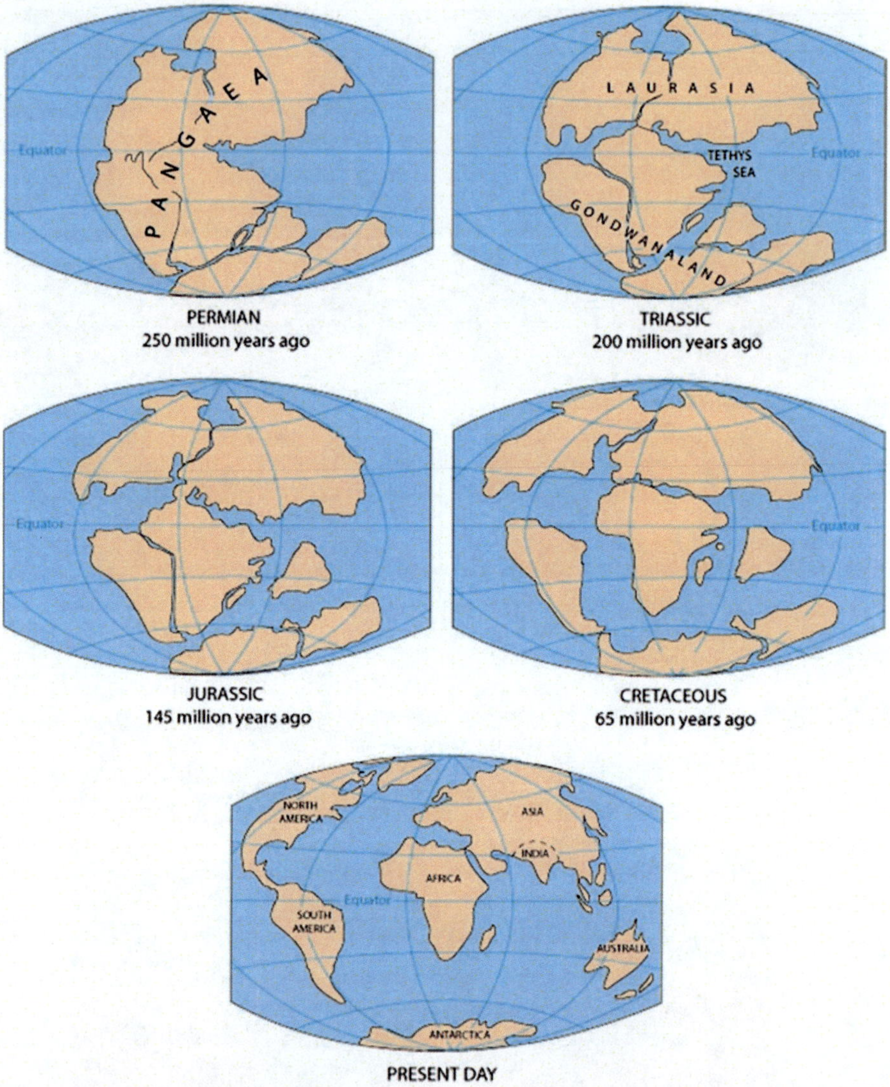

Fig. 8.3 Continental drift over geological time

combined in a 'supercontinent' called Pangea which, because of plate tectonics, reassembled in different patterns until achieving the appearance we currently view. But changes will continue to occur over the many millions of years the core of earth yet remains molten.

Meanwhile, the present configuration of tectonic plates has created a substantial area of earth known as "the ring of fire" that is highly prone to earthquakes and, where plate subduction occurs, volcanos may appear (Fig. 8.4). Note that "the ring

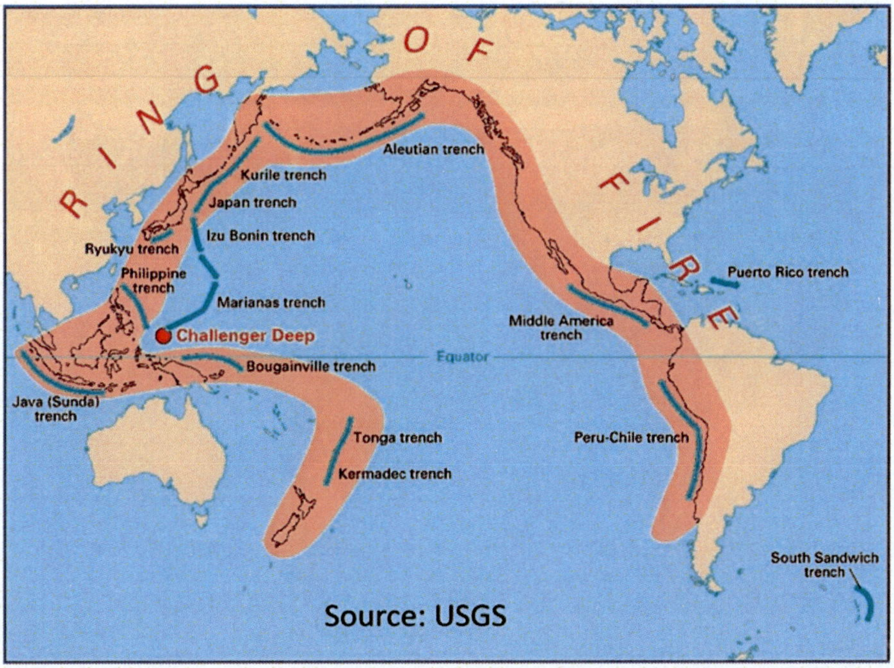

Fig. 8.4 Ring of fire

of fire" encompasses a good portion of the west and eastern Pacific, including western Canada and the United States as well as Western South America.

It might appear at first glance that it is only the western coast of America is subject to earthquakes, but that is not so. The United States Geological Survey maintains constant surveillance of those factors that influence the frequency and magnitude of U.S. earthquakes and based on their data, create seismic hazard assessment maps.

> Earthquakes cause an estimated annualized loss to the U.S. of several billions of dollars. To mitigate earthquake losses, it is necessary to evaluate the earthquake hazards across the country. The seismic hazard maps address this need by integrating what scientists have learned about earthquake sources, crustal deformation, active faulting, and ground shaking. This information is translated into a form that can be used to reduce the risk from earthquakes and to improve public safety. The resulting seismic hazard maps are improved and updated on a periodic basis by incorporating new information. The USGS maps are the basis for seismic provisions in building codes and for risk models used in insurance rate structures. An integral part of this project is a database describing Quaternary faults and digital maps of those faults for the U.S. and its territories. (USGS, 2014)

Figure 8.5 is an example of such a map. I think most Americans would be surprised to see that a large portion of the U.S. has a moderate (yellow) to high (red) susceptibility to earthquake–hazard … even more surprised to see the number of

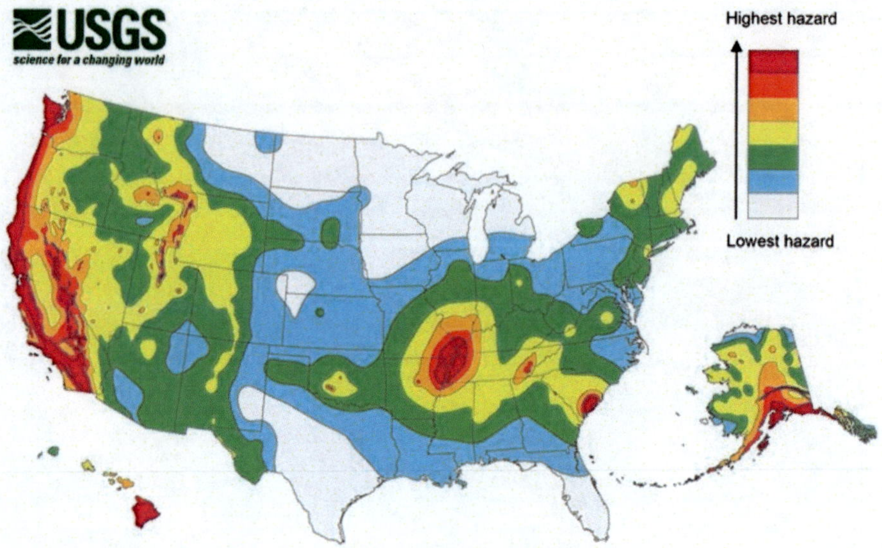

Fig. 8.5 2018 Long-term national seismic hazard map

Table 8.1 Earthquake magnitude scale

Magnitude	Earthquake effects	Estimated number each year
2.5 or less	Usually not felt, but can be recorded by seismograph.	Millions
2.5 to 5.4	Often felt, but only causes minor damage.	500,000
5.5 to 6.0	Slight damage to buildings and other structures.	350
6.1 to 6.9	May cause a lot of damage in very populated areas.	100
7.0 to 7.9	Major earthquake. Serious damage.	10-15
8.0 or greater	Great earthquake. Can totally destroy communities near the epicenter.	

earthquakes within States and Puerto Rico (an unincorporated territory of the United States) on a single day.[2]

It is true of course that, from a global perspective, many earthquakes are relatively mild, as shown in Table 8.1, but well over a million quakes per year cause substantial damage.

Over 100 of the largest quakes in the United States in the period 1929–1946 were of magnitudes that ranged from M7.0 to M8.2 which, in English, means 'major' to 'great' earthquakes; of the 16 largest in the period 1952–2014 (USGS), magnitudes ranged from M7.3 to M9.2, meaning 'major' to, I suppose, 'very great' *(National Earthquake Information Center, U.S. Geological Survey).*

[2] See Appendix 4.

I am writing this paragraph 6 days after the August 14, 2021-earthquake in Haiti. The magnitude is known to have been M7.2, well within the category of "major and serious danger". A few days after the quake, the death toll "hovered" around 2000. However, based on models of hurricanes in heavily populated areas, the USGS (United States Geological Survey) suggested that the actual number of fatalities would likely fall within the range of 10,000 to 100,000 or more, and that the total monetary cost would likely be somewhere between $100 million and $14 billion (Reardon, 2021). Five days after the quake, it was reported that at least 12,000 people were injured, and 600,000 in need of emergency assistance (BBC News, August 19, '21). On the same day, UNICEF (United Nations International Children's Emergency Fund) announced that 500,000 children had no access to shelter, safe water, and food.

Well, so what? Just why have I spent the last several pages on a rather elementary discourse on tectonic dynamics and conclude with a linguistic snapshot of the horrors of the earthquakes those dynamics produce? Afterall, humans have no control of tectonic phenomena. However, as I emphasized at the beginning of this chapter, my intent is to emphasize key interactions between human behavior and global dynamic processes, and how human behavior must be considered in comprehensive and effective emergency planning and response.

Toward that end, the remainder of this chapter (as well as the next) is devoted to the following two considerations:

1. Given the horrific costs in lives, physical and psychological injuries, personal economics, social upheaval, and just plain misery and despair, how do we measure the effectiveness of amelioration efforts in terms of the *duration of response,* its *equitable application*, and *comprehensiveness regarding diverse expertise….* in short, its *humane design and inclusiveness (for both immediate and generational victims as well as subsequent responders)?*

I think it appropriate to emphasize the meaning of "humane", a word that is typically not used in daily conversation: it is *not simply the expression* of sympathy and consideration for others (which is quite easy to do), *but acting accordingly*, which can often require significant cost (however 'cost' may be measured).

2. While we do not have any control over tectonic dynamics, we Americans can have some influence on human behavior regarding '*land-use control*', … that is, where we may or may not reside, activities allowed or disallowed in designated areas, and the allowed design of structures in which we live and work.

Regarding the legality of land-use control in the U.S., the Disaster Mitigation Act of 2000 requires states to create and submit plans regulating development in disaster prone areas. State compliance is expected to include major efforts to ensure local municipal participation in the planning process and is rewarded with federal disaster relief payments. One reason for the emphasis on local participation is the simple fact that local economics and environments are quite different from state-wide and/or national financial and environmental concerns; obviously, local authorities are much more familiar with local circumstances and have a personal stake in economic success and meeting local expectations regarding the quality of life. Given the wide variety of state, local, and national objectives, national legislation regarding land-use control, legal system must remain open to the spirit of on-the-ground experimentation rather than of subservient obeisance to a top-down command (Nolon, 2006).

From the early nineteenth century, Americans have benefited greatly through the progress of science and subsequent invention, particularly in technology, but not necessarily in other areas of social concern such as land-use control. Being now almost a quarter of way through the twenty-first century, I suggest we pay significant attention to how best to formulate and implement a thoroughly modern approach to emergency-prevention, which logically raises the question "why do we not only allow but even facilitate (e.g., granting mortgages for homes; loans to businesses) people moving into earthquake-prone areas?". I know … the question sounds very un-American. But I stubbornly think it a logical question.

Invention, mind you, evolves out of repeated experimentation. Well, … in America we have 50 States in which we can conduct the appropriate experiments. We've done it before. We can do it again…. or do you think experiments should be confined to a laboratory, and not attempted in those political-social constructs we call 'States'?

8.2 More to the Point

Figure 8.6 is (once again) a modification of previous Figs. I've included in earlier chapters in order to highlight two facts: (1) that earthquakes occur not only because tectonic pressures produce 'breaks' (faults)in continental plates, but also in oceanic plates, and (2) that humans are not only victims of both types of quakes but, based on an ever increasing understanding of quake-dynamics, humans can also contribute to various ground-fault failures on the land-surface.

The USGS identifies quake hazards on land as follows: *Ground Shaking* (or ground vibration) that varies with increasing magnitude and decreases with distance from the causative fault; *Ground Failure*, as due to *liquefaction*, when soils temporarily lose strength and behave as viscous fluids, and lateral movements ('lateral spreads')of large blocks of soil as a result of liquefaction,[3] which are particularly destructive of pipelines; *Flow Failures*, as liquified or blocks of soil or other solid material riding on a layer of liquefied soil (on both land and underwater); *Loss of Bearing Strength*, as when soil supporting a structure loses strength (due to liquefaction) and tips the overlying structure; and *Landslides*, as rock-falls and slides of rock- fragments that form on steep slopes.

[3] Note that liquefaction can occur because of rainfall on soils de-vegetated by wildfires (Chap. 4); a good example of how one emergency (earthquake) can magnify the consequences of a previous emergency (wildfire).

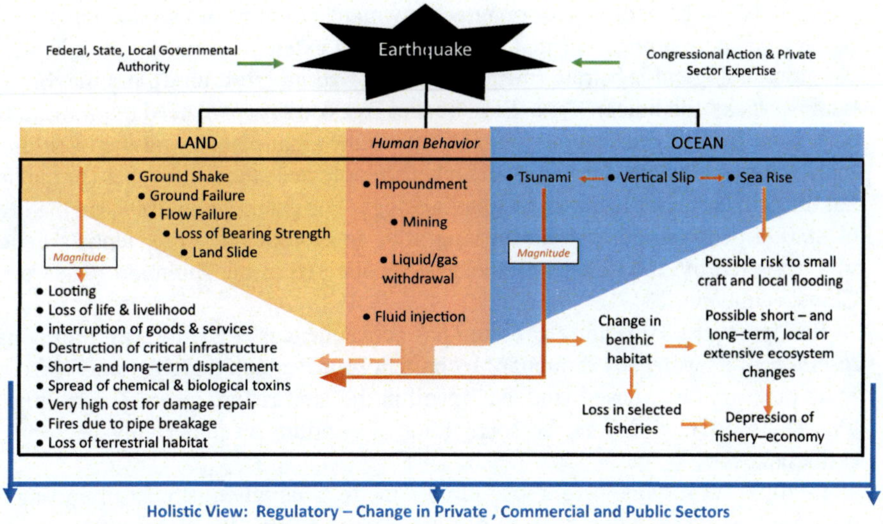

Fig. 8.6 Interconnections of causal chains associated with earthquakes

While humans have no control over plate tectonics, we can and do influence quakes in some situations. For example, the USGS has evaluated human activities of particular interest, including liquid/gas withdrawal from sub-surface reserves, fluid injection into geologic-strata, mining, and impoundment of surface water-supplies (i.e., reservoirs).[4]

Fracking, a process using pressure on oil and gas wells goes back to mid -1800s. Almost two million U.S. wells have used this process to produce billions of barrels of oil and trillions of cubic feet of natural gas. In the 1980s, concerns regarding the environmental effects of fracking became controversial, with New York even banning its use. Based on its study of the effects of fracking in Oklahoma, which has the most induced earthquakes in the U.S, the USGS has concluded that fracking causes only 1–2% of induced-earthquakes, and that the rest are primarily caused by well-wastewater (fluid)-injection. However, most injection wells are not associated with so-called 'felt earthquakes' which result from a variety of factors, including the presence of underground faults.

The effects of reservoir impoundments on earthquakes (known as 'Reservoir-Triggered Seismicity', or RTS) were first reported in the U.S.in 1935 as related to operations at the Hoover Dam and have since been observed in various other global and American locations. Among a variety of complex factors playing a human-induced quakes, two major factors include (1) the weight of impounded water, and (2) the seepage of water into underground faults (cracks 'n holes). As I stressed earlier, pressure has consequences.

Consider the weight of water. One cubic foot of water weighs 62.4 lbs. Imagine a 1-cubical-ft- column of water that is 50 ft. long; the weight of that single column

[4] USGS, Earthquake Hazards, Myths and Misconceptions about Induced Earthquakes.

is 3120 lbs., or 1.56 tons. Now imagine how many columns of variable length combine to make up a reservoir that extends over an extensive expanse of land with a variable geological substrate (different kinds of strata [soft, hard] and maybe containing cracks and holes). Voila. Earthquake! But you say, "We need the water". Yes, of course. But .... did you put the reservoir in the right place? And then you might reply, "Hey, we have to put reservoirs where people need the water". Yes (or maybe), but the critical factor is the underlying geology. For example, in 2008, on the order of 80,000 people died or went missing after an earthquake that scientists believe was triggered by 320 million tons of water collected in the Zipingpu Reservoir in China. (Gibbons, 2017).

Really, must I say again, "Pressure has consequences"? By the way, emptying a reservoir of its water (by drought or by human contrivance), thereby removing overlying pressure on exposed land, can result in the land rising, just as the melting of glaciers (see Chap. 9 does, both resulting (depending on underlying geology) in earthquakes.

In 2007, six miners were trapped in a mine in Utah when beams supporting the mine roof ruptured. After 4 weeks of futile rescue efforts, the mine was sealed and declared an M3.9 earthquake catastrophe. However, after extensive seismological study, it was determined that the earthquake was not the cause of the mine-collapse, but rather, the collapse caused the earthquake. Since then, Mine-Induced-Seismicity (MIS), has become the vernacular replacement for "bumps and bounces". Such MIS quakes are very often of small magnitude (below M3.0), but, depending on geology and depth of mine, have reached magnitudes of M4-M5 in other locations wherever mining is conducted (Cook, 2015).

Turning our attention to oceanic quakes, it's probably useful to use a simple analogy. For example, imagine yourself in a bathtub and holding a large plate on your lap. Now raise that plate abruptly. Depending on the force you exert, you can make very small and somewhat larger waves at the surface. Of course, the analogy would be better were your bathtub engineered with a bottom that could be mechanically raised and lowered, but I haven't seen one of those.

Depending on the magnitude of the oceanic fault (the pressure of upthrust), its frequency and duration, and the depth of ocean, you might have a small, temporary and localized rising of water; under extremes of the relevant factors, you get a Tsunami. Where the ocean is several miles deep, Tsunami-waves can travel more than 400 miles per hour; as they approach land, their height increases greatly up to as much as 80 or more feet (USGS, Earthquake Hazards).

Just imagine standing on the seashore and suddenly seeing an 80-foot-high wave moving faster than a racecar toward you! Between 1998 and 2017, Global tsunamis killed more than 250,000 humans. Reconstruction costs of a single tsunami may cost from several billion to, as in the case of the Fukushima tsunami, several hundred billion (thus far). Some of the most catastrophic Tsunamis are listed in Table 8.2.[5]

[5] Note that the 'Japan 2011' Tsunami entry in Table 8.2 is responsible for the infamous nuclear disaster in Fukushima on March 11, 2011. This is yet another example of how one emergency (earthquake) results in a potential future emergency event (i.e., the disposal of radioactive waters into the ocean proposed 10 years after the tsunami).

Table 8.2 Examples of catastrophic Tsunamis

Country	Year	Magnitude	Wave Height (feet)[a]
Chili	960	9.5	32–82
Indonesia	2004	9.1–9.3	49–167
Chili	2010	8.8	6.5–39.4
Japan	2011	9.0–9.1	132.9
Indonesia	2018	6.4–7.4	6.7–16
Indonesia	2018	7.5	19.7
Portugal	2018	7.1–7.3	19.7–98.4

[a]Range depending on distance from epicenter

Tsunamis also have important impacts on oceanic (as well as terrestrial) habitats. Wave erosion of bottom sediments and subsequent deposition in other areas change the physical basis of habitats, with significant effects on both vertebrates and invertebrates, depending on the amount, physical characteristics (e.g., granularity) and chemical quality (e.g., nutrient content and toxicity) of sediment. It is typically impossible to forecast the actual effects of these sediments, either on terrestrial or aquatic ecosystems; we must wait until after the tsunami to analyze the transported sediment, particularly given the land-based contaminants that will be backwashed into the ocean, and to determine the inland-extent of a tsunami-wave into urban, agricultural, and otherwise vegetated land. It is nonetheless evident that the transport of sediment as well as chemical contaminants from and into terrestrial and aquatic systems may result in significant economic impacts.

8.3 Overview of Fig. 8.6: Holistic Response

Table 8.3 includes a list of 15 American cities that the U.S. Geological Survey classifies as *most prone to a 'Devasting Earthquake'*. I have included the year each was founded as well as their population as of 2020 to emphasize the fact that that, for any of the innumerable reasons for human migration, each has experienced a persistent and even geometric increase in population from their founding to 2020. For example: Memphis grew from ~40,000 (*1870*) to ~600,000; Los Angeles, from ~2000 (*1790*) to ~10,000,000; San Francisco, from ~1000 (*1840*) to ~4,700,000; New York City, from ~33,000 (*1790)* to ~8,800,000; Washington D.C., from ~13,000 (*1820*) to ~700,000.

Please excuse my imagery, but looking at these numbers, I am sorely tempted to envision a herd of buffalo stampeding toward a cliff!

Of course, cities are extremely important, necessary, and to many millions, most desirable. And, yes, we have made much progress in the architectural design of structures to withstand or at least lessen earthquake damage; much less so regarding the ravages of Tsunamis.

Table 8.3 Ten most earthquake-prone U.S. cities

U.S. cities most prone to devastating earthquake[a]	Year founded	2020 population
Memphis	1819	633,104
Los Angeles	1781	10,014,009
San Francisco	1776	4,690,000
Seattle	1851	737,015
Oklahoma City	1889	681,054
Anchorage	1914	291.247
Tulsa	1895	413,066
Dallas	1841	1,304,379
Portland	1851	652,503
Salt Lake City	1847	199,723
Honolulu	1907	350,964
Wichita	1864	397,532
New York City	1624	8,804,190
Charleston	1670	408,235
Washington D.C.	1790	689,545
Total		*30,266,566*

[a]based on USGS data

It has been estimated that, by 2050, 75% of future Americans will live in cities and urbanized areas, which is to say the same number of Americans living in cities as there are in the entire nation today (University of Michigan, 2019). That fact to one side, it is important to consider not only urban, but also suburban and rural (combining the three into.

landscape) populations with respect to quake- emergency planning and response, especially regarding local topographical susceptibility to Tsunami-flooding.

After all, "most of the densely populated countries and cities lie on the Ring of Fire.[6] Lima, Quito, and Santiago in South America; San Diego, Los Angeles, and San Francisco in North America. Then you have whole countries with large and concentrated populations, such as Japan [127 m people], Philippines [103 m people] and Indonesia [267 m people] "(*IRG, Integral Risk Global*).

Obviously, the magnitude of any quake-related emergency is directly proportional not only to the magnitude of the quake, but also to the density of the urban human population. About 64% of the U.S. population lived in urban areas in 1950; that will likely increase to 89% by 2050 when 63% of the world population will also be living in cities (University of Michigan, 2017*)*. However, the magnitude of quake-related emergency is also directly proportional to the social inequities within that population.

[6]Refer back to Fig. 8.4.

For example, consider the 7.2 magnitude earthquake in poverty-stricken Haiti in 2021 in which 50,000 homes were destroyed and 77,000 severely damaged, 60 places of worship, 20 schools, and 25 health centers were destroyed or damaged and 48 foster homes that care for 1700 children were damaged only eleven years after an earthquake that left 1.5 million Haitians homeless (Reid, 2021).

Well, Haiti is among the poorest nations of the world. The United States is the richest, yet we also have many poor (and otherwise disadvantaged by various inequities) who, in any particular emergency, suffer significantly more than their financially and socially advantaged fellow Americans.

Most recently, we have seen this in various consequences of the COVID pandemic; we also have strong hints of it regarding emergency response efforts during Hurricane Katrina in 2005 that "*sparked renewed debate about race, class, and institutional approaches toward vulnerable population groups in the United States. In the aftermath of the storm, a wide array of media reports, public statements, and polls underscored this reality*" (Select Bipartisan Committee, 2006).

In the above quote, the phrase 'institutional approaches toward vulnerable population groups' is a highly politicized expression that means, in academic-English, "institutional inequities" which, in turn, translates most directly into "unfair but legal prejudice" that includes color of skin, income, level of education, gender, ethnicity, and all such other excuses we humans use to elevate ourselves above others.

As I write, I am mindful of pleas of a previous author (Cutter, 2001) regarding necessary improvements that should be made in emergency planning and response … improvements that I summarize needed improvements yet to implemented:

- The need to integrate social systems, environmental systems, and the built environment
- The need to identify certain segments of society that are disproportionately affected
- The need to achieve the fair distribution of justice

Well, those needs were identified over 20 years ago and, unfortunately, have been largely ignored. The pertinent question is: "Why has it taken almost a generation to begin even to consider giving serious attention to their implementation?"

Quoting Mark Twain (again), "I dunno", but I'm pretty sure it has to do with how decisions are made (or not made) in our highly complex, politicized society. I will discuss this further in Part III of this volume. Meanwhile I ask that you keep this question very much in mind as you read Chap. 9.

Suggested Exercises for this Chapter

1. Using the internet, download a map of the United States showing major mining areas and superimpose a measure of population densities in those areas. Write an expository essay (and/or give an oral presentation) on your reactions to what you see.
2. Alternative to question #1: use a map of the "Ring of Fire" and superimpose population densities.

3. Choosing among the following examples of social inequities that are of current concern in America, give an oral presentation (or essay) of key examples and explain how they are relevant to your generation.
4. Given (A) that there are significant regional differences in the U.S. regarding earthquake vulnerability (including ground-shake and/or tsunami), and (B) that many of these regions include densely populated cities, identify, and discuss possible steps that might be taken to ameliorate disaster.

References

Cook, T. (2015, April 3). Ground-shaking research: How humans trigger earthquakes. *EARTH*.

Cutter, S. L. (2001). Charting a course for the next two decades. In S. L. Cutter (Ed.), *American hazardscapes: The regionalization of hazards and disasters*.

FEMA. (2017, June). *National earthquake loss estimate*.

Fernandez-Armesto, F. (2001). *Civilizations: Culture, ambition, and the transformation of nature*. The Free Press.

Gibbons, S. (2017, October 2). *How humans are causing deadly earthquakes*. National Geographic.

Nolon, J. R. (2006, September 8). *A diagnostic approach to evaluating governmental land use control*. Pace University, Pace Law Faculty Publication (Article 8).

Reardon, S. (2021, August 19). The true Haiti earthquake death toll is much worse than early official counts. *Scientific American*.

Reid, K. (2021, August 23). *Haiti earthquake: Facts, FAQs, and how to help*. World Vision.

Select Bipartisan Committee. (2006). *A failure of initiative: Final report*. U.S. Government Printing Office.

Simpson, D. W., et al. (1981). *The 1906 San Francisco earthquake and the seismic cycle, earthquake prediction: An international review* (Maurice Ewing series 4, pp. 126–140). American Geophysical Union.

University of Michigan. (2017). Center for Sustainable Systems (CSS). U.S. Cities Factsheet.

University of Michigan. (2019). Center for Sustainable Systems (CSS). U.S. Cities Factsheet.

USGS (United States Geological Survey). (2014). *How would sea level change if all glaciers melted*.

Chapter 9
Climate Change

Our earth is about 4.5 billion years old and, based on our current scientific understanding, the earliest life on earth appeared 3.7 billion years ago. Such measures of time are meaningless in our daily lives wherein our personal experience over the various summations of minutes, hours, weeks, and months finally add up to less than a mere 100 years. Considering the many changes we individually experience in our short lifetimes, we have to be in awe of the changes that earth itself has experienced as it has evolved from a molten mass to a water-world; from a sphere of ice to a world of oceans dotted with newborn continents; from barren land to lush forests; from tropical biomes to ice ages interspersed over millennia with thousand-year thaw.

In short, from its birth, the earth has been in a constant state of change, and not only in its physical (and, therefore, its biological) characteristics; even time itself has changed. At its earliest, the length of an earth-day was 18 and not the current 24 h.

So what is all the contemporary fuss regarding 'climate change' about? Simple! For the first time in earth's 4.5 billion years, climate is changing due not only to the various factors produced and long-employed by Nature, but by human-production of greenhouse gases (GGs).

Such gasses absorb infrared radiation reflected by the sun-heated earth, transfer that heat to the atmosphere which, in turn, transfers it back to earth. In short, they trap heat on earth.

Major human-produced GGs include (in order or relative amounts):

Carbon Dioxide (CO_2) 75% *from* burning of fossil fuels and wood
Methane (CH_4) 14% *from* agriculture (livestock and rice)
Nitrous Oxide (N_2O) 8% *from* agriculture (fertilizer) & industrial processes
Fluorinated gases 1%. *From* variety of industrial processes (The Guardian, 2011).

Note that Fig. 9.1 shows pretty much of a steady increase in earth's temperature from the beginning of the Industrial revolution (1800) to the present, and a clearly consistent linear increase starting in the mid 1960s a pattern that directly reflects the rampant industrial expansion in mid-twentieth-century America that yet continues into the twenty-first.

P. A. Erickson, *Effective Environmental Emergency Responses*,
https://doi.org/10.1007/978-3-031-05893-6_9

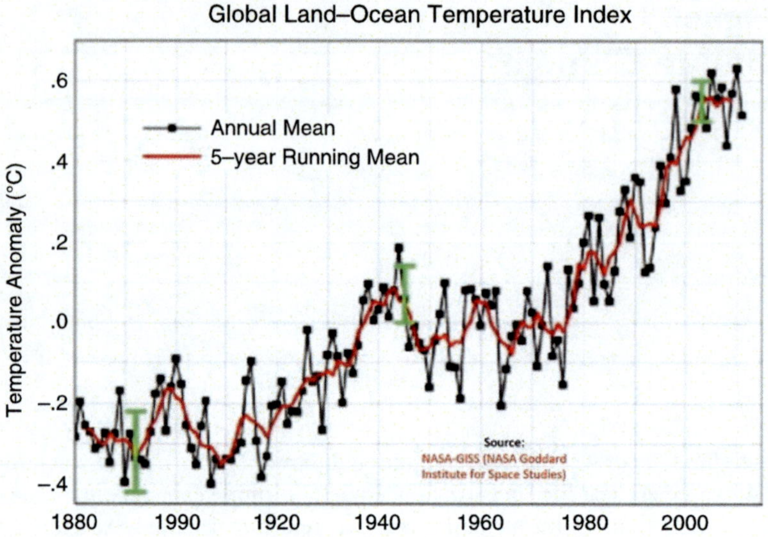

Fig. 9.1 Global land-ocean temperature index

Despite the naysayers among the general public as well as those with (maybe) some science but, nonetheless, with abundant zest for promulgating misinformation and stubborn denial, the world at large has (finally) come to agreement that the earth has been warming up measurably ever since the industrial revolution and that humans (and not Nature) are largely responsible. Endorsed by the UN General Assembly, the Intergovernmental Panel on Climate Change (IPCC), comprised of 195 member states and aided by literally thousands of scientists world-wide, has issued a recent report that contains clearly alarming conclusions regarding thermal consequences of human contributions to climate change, including:

- "unless there are immediate, rapid and large-scale reductions in greenhouse gas emissions, limiting warming to close to 1.5 °C (34.7 °F) or even 2 °C (35.6 °F) will be beyond reach."
- "emissions of greenhouse gases from human activities are responsible for approximately 1.1 °C (33.98 °F) of warming since 1850–1900 and, averaged over the next 20 years, global temperature is expected to reach or exceed 1.5 °C (34.7 °F) of warming."
- "in the coming decades climate changes will increase in all regions. For 1.5 °C (32.7 °F) of global warming, heat extremes would more often reach critical tolerance thresholds for agriculture and health."
- "there is alarming evidence that important *tipping points*,[1] leading to irreversible changes in major ecosystems and the planetary climate system, may already have

[1] Ramirez, R., We are "73% of the way to 1.5 °C threshold scientists warn the planet must stay under to avoid the worst implacts" (CNN, January 10, 2022).

been reached or passed. Ecosystems as diverse as the Amazon rainforest and the Arctic tundra, may be approaching thresholds of dramatic change trough warming and drying. Mountain glaciers are in alarming retreat and the downstream effects of reduced water supply in the driest months will have repercussions that transcend generations (IPCC, 2021)"

Please take special notice that I have underlined the phrase 'tipping point' in the last bullet above. It's a simple concept but extremely important. For example, imagine that you have a container filled with a precious material such as gold-dust. Unfortunately, you tip that container into the wind which, of course, scatters the gold-dust. Well, you can't get it back again. In any emergency we often hear the phrase "return to normal". Well, sometimes, there is no return. Like that gold-dust, what we consider to be normality can utterly disappear and be replaced by a totally new normality. That is what climate change is all about.

While the sun is roughly 5 billion years old it still has roughly another 5 billion years to go. That's a long enough time that we don't have to worry about the sun's output of heat; that will remain rather constant for a very long time. However, our climate is also determined by how much of the sun's heat we humans trap on earth.

The following conclusions of various studies clearly show that humans are very efficient "thermal trappers":

- The Arctic is among the fastest warming regions in the world and is heating more than twice the global average (World Meteorological Organization, 2021).
- Lakes in the Northern Hemisphere are warming six times faster than any other time-period in the last 100 years (York University, 2021).
- Himalayan glaciers have lost ice ten times more quickly over the last few decades than on average since the last major glacier expansion 400–700 years ago, (University of Leeds, 2021
- Ice scientists meeting in New Orleans warned that something even more alarming was brewing on the West Antarctic ice sheet – a vast basin of ice on the Antarctic peninsula. Years of research by teams of British and American researchers showed that great cracks and fissures had opened both on top of and underneath the Thwaites glacier (The Guardian, 2021)

Yes, it is clear that sea-ice as well as glaciers are melting (Fig. 9.2) which results, in turn, in rising sea levels (Fig. 9.3). How much? In the extreme (i.e., all melting occurs at the same time), ocean levels would rise about 230 ft and flood all coastal cities *(USGS)*. More realistically, melting will occur over time, which raises the question "how long will it take?". The following facts first published in 2020 (and updated December 21, 2021) may well suggest that the future is fast-approaching.

Glacial melting results is raised sea levels ......over the last 140 years, just over 8 inches; in the last 30 years, almost 4 inches, which shows that the rise in sea level is accelerating……from 0.06 inches throughout most of the twentieth Century to 0.14 inches from 2006–2015 (Lindsey, 2020–2021).

Of course, the impact of sea level rise varies with location and topological nature of shoreline. For example, Fig. 9.4 is an example of sea-rise effects along the east

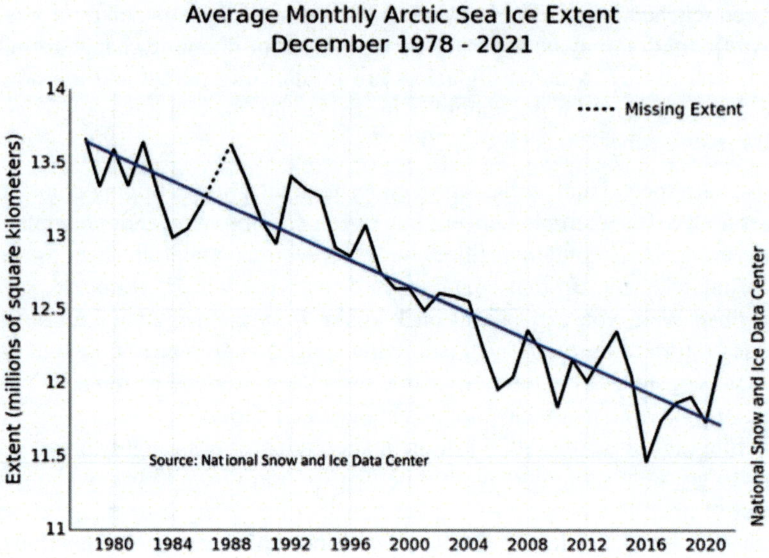

Fig. 9.2 Sea Level Rise

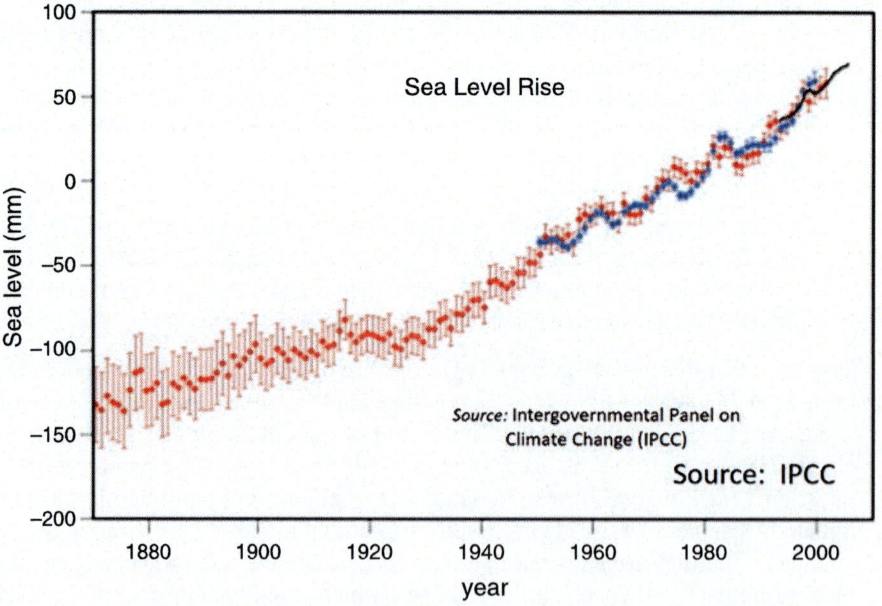

Fig. 9.3 Average monthly Arctic sea ice extent December 1978–2021

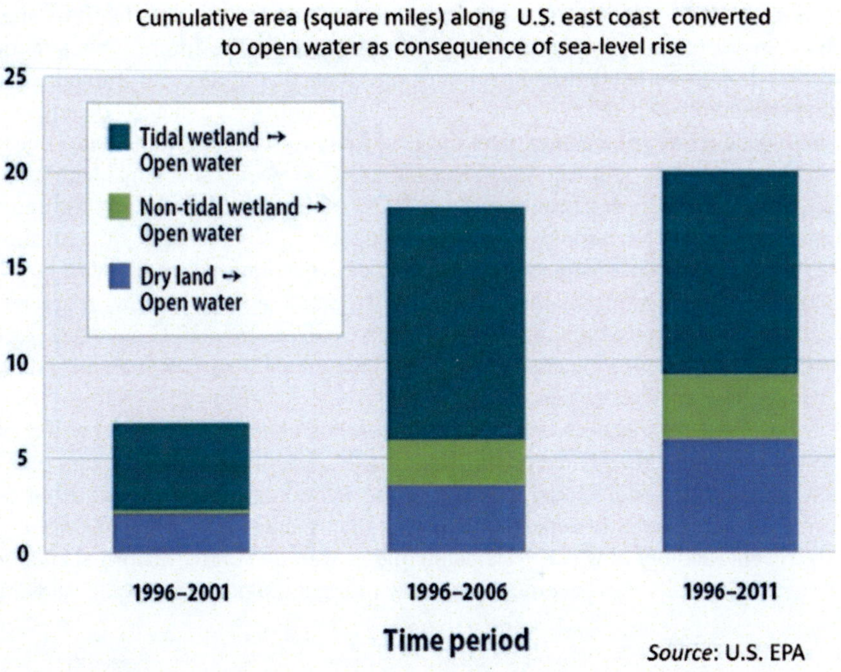

Fig. 9.4 Cumulative area (square miles) along U.S. east coast converted to open water as consequence of sea-level rise

coast of the U.S. over the period 1996–2011. But this is only one example from one geographic region. There are many other coastal land areas throughout the world[2] that could be completely under water by 2050, including:

Adriatic Coast, Italy	Alexandria, Egypt	Andalucia, Spain
Bangkok, Thailand	Bruges, Belgium.	Charleston, South Carolina, USA
Delft, the Netherlands	East Anglia, UK	Everglades National Park, Florida, USA
Florida Keys, USA.	Ho Chi Minh City, Vietnam	Hudson Bay, Canada
Hue, Vietnam.	Kolkata, India.	Kuching, Malaysia
London, UK	Manila, Philippines	Mumbai, India
New Orleans, Louisiana, USA.	New York City, New York, USA	Panama City, Panama
Port Douglas, Australia.	Ribe, Denmark	San Francisco Bay Area, California, USA
	Shanghai, China	

[2] see https://www.loveexploring.com/gallerylist/91927

Given the ever-increasing complexity of our daily lives, it is difficult for many to have much (or any) concern over events forecast for the future. Our attention is focused on those hectic minutes, hours, and days that dictate our immediate needs and obligations. The terms "short-termism" and "temporal exhaustion"[3] are increasingly used to describe and explain our proclivity to focus on immediate necessities and opportunities and not future problems. One need only consider the severe upheaval in our lives caused by the COVID pandemic that abruptly shifted daily attention away from our usual daily preoccupations to personal health and survival.

The Climatic data collected over almost nine generations are likewise unmistakable. The previously hypothetical future of climate change is now. If we take no action now to avoid a climatic tipping point[4], a baby born in 2022 anywhere on earth will daily experience the chaos of climate change within a few years of graduating from college or even earlier.

Despite the science-based surety of impending chaos, there is (and will continue to be) much denial and outright refusal to take appropriate action by very real institutional and personal vested interests in the technologies and natural resources on which we are currently dependent. On the other hand, as demonstrated not only by the recent (and ongoing) COVID pandemic as well by similar historical challenges, extreme adversity can become the nursery of technological and social innovation.

9.1 A Holistic View of Climate Change

Throughout this volume I have used the phrase "interconnected causal chains" in my discussions of the various consequences of certain types of emergencies. Since 2011, other phrases have been widely used by academic and practicing emergency-related professionals, including "cascade analysis", which most often connotates "toppling dominoes" or "the bubbling cascades of waterfalls".

Either metaphor clearly emphasizes the both the *knowns* and *unknowns* in any complex system. How does the density of dominoes influence their spacing to achieve a toppling? what If the length of different dominoes varies? How does temperature of water or the frictional value of stones or wind affect waterfall spray? There are many variables that make such metaphors rather vague and, therefore, result in various definitions of "cascade" (Pescaroli & Alexander, 2015).

Science always deals with unknowns, and it most often takes a long time for science to invent the known. For example, it took several hundred years for Einstein finally to supplant Newton's concept of gravity. While I have due respect of the concept of "cascade" as a useful tool in emergency planning, I believe that, given our experience with actual emergencies and our persistent efforts to understand the

[3] Both terms basically describe a personal preoccupation with existing problems to the detriment of concerns regarding the likely problems of future generations.

[4] I emphasize that a tipping point indicates that a change from one condition to another cannot be reversed.

underlying sciences of those events, we do have the knowledge to identify "interconnected causal chains" *and act accordingly*.

The key issue regarding emergency planning and response is not the lack of knowledge, but rather our failure to make appropriate decisions. I learned that lesson from the U.S. Fire Administration which has long examined each fire-event to determine not only what was 'done-right', but more importantly, what was 'done-wrong'. Thus, in the following discussions of climate change, I give particular attention to decision-making which is the primary focus of Part III of this volume.

Given the complexity of interconnected causal chains in the phenomenon of climate change, I constructed Fig. 9.5 as a series of concentric roulette wheels. Each wheel revolves independently and, at any time in either direction with varying speeds. Please consider phrases in each wheel to be equivalent to 'slots' and arrows to be the 'ball' in a real roulette game.

We start with the center (green) wheel representing both Nature's and Human contribution to Climate Change. As I've already pointed out, the Human production of greenhouse gases has been increasing the speed with which this wheel turns ever since the early 1800s (beginning of the Industrial Revolution) while the sun's output of heat has remained essentially constant over billions of years. Moving to the next wheel, we see the temperature on earth increasing and, to the outer wheel, glacial-melt and subsequent sea-rise that leads to coastal flooding that, in turn will certainly result in massive human migrations. Those high temperature which are already

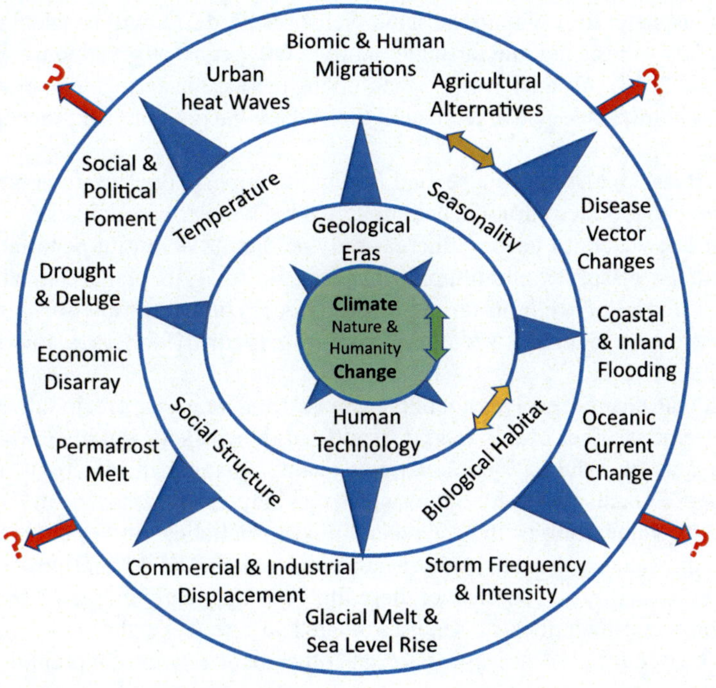

Fig. 9.5 Consequences of climate change

melting glaciers and sea ice will also melt the permafrost which will, in turn, permit the bacterial decomposition of vast amounts of otherwise frozen organic material with consequential release of untold tons of carbon dioxide to entrap even more heat. Such carbon dioxide, added to the pool produced by human technology will also increase the production of atmospheric carbonic acid which as rainfall has long increased the acidity of our surface waters.

Depending on region, increased temperature also changes the time-frames of seasons. Winters can come later; springs and summers can come earlier Temperature also influences continental storms and droughts as well as ocean currents that are so vital to the distribution of global heat. Seasonal changes, inland drought and deluges are devastating for agricultural crops. Drought alone is often the prelude to wildfires which, in the recent few years, have consumes many billions of dollars and shattered many lives.

Wildfires, droughts, and seasonal and high temperatures will have severe impact on biological habitat, resulting in wildlife migrations that can spread disease.

Well, the above are examples of what is already in progress and well documented in daily newscasts throughout the world. However, I think it useful to consider several of these phenomena in more detail.

1. For example, permafrost melt will result in the exposure of thick deposits of organic material (vegetative and animal remains) that have been frozen for millennia. Once they are defrosted, these remains will undergo decomposition in less than a year with the release of gigatons (i.e., billions of tons) of CO_2 to the atmosphere; fires will do the same. Some released CO_2 will be taken up by plant photosynthesis, but the far-larger amount will persist as greenhouse gas (Schuur et al., 2013). As such, it will speed-up the increase in earth's temperature, a process known as 'positive feedback' (i.e., where the product of a process speeds up that same process.

   But CO_2 also plays a second role in the atmosphere; it interacts with water vapor to produce carbonic acid that falls back to earth as acid-rain, causing our surface waters to become increasingly acidic. As acid rain penetrates mineral soils, it dissolves aluminum which can be toxic to plants and animals and therefore produce long-term changes in ecosystems. It took 50 years of intensive study to finally recognize acid-rain in the 1980_s as a major problem (EPA, 2020).
2. Climate change involves not only increased temperatures, precipitation extremes, extreme weather events (e.g., tornadoes & hurricanes), but also produces poor air quality, reduced food and water quality, changes in infectious agents, and population displacement, all of which can play significant roles in local human-health vulnerabilities to heat-related illness, including cardiopulmonary illness, food-, water- and/or vector-borne disease, or mental health and stress. Of course, the probability and severity of these illnesses are greatly increased by the persistent social inequities in America related to age & gender, color & ethnicity, poverty, education, access to care, and other manifestations of prejudice (Climate Central, 2021).

I should point out that the proposed relationship between climate change and subsequent biotic migrations that might spread vector-borne infectious diseases was controversial 10 years ago. After all, ecological systems are extremely complex and ecologists considered that there was insufficient evidence to support the proposed hypothesis (Rohr et al., 2011). Climate change will result in animal migrations, but it is also true that it is the combination of human behavior and exposure to infectious vectors that determines vulnerability to illness (National Center for Environmental Health, 2020).

As I have said before, Science takes time. Meanwhile, the U.S. EPA states, "Climate change is expected to affect the geographic and seasonal patterns of vector-borne diseases (that is, diseases caused by pathogens transmitted by mosquitoes, ticks, fleas, or other arthropods) in the United States" (EPA, 2021a, b).

It is reported most recently that, because the warming of water results in a decreased concentration of oxygen, it is expected that by about 70% of the world's oceans could be so deficient in oxygen as to impact marine ecosystems world-wide. Models of marine dynamics indicate that, having passed a critical threshold in 2021, many fisheries are increasingly experiencing oxygen-deficiency (Science News, 2022a, b).

3. Heat, drought and deluge, floods, hurricanes, wildfires, and altered seasonal changes will certainly present serious challenges to both agriculture and husbandry. The U.S. Department of Agriculture reports the following:

   - The 10 warmest years on record occurred in the years 2005–2020
   - Droughts, floods, and hurricanes in 2020 totaled at least $1 billion, more than any other year since 1980
   - Six states in the U.S. had the lowest precipitation and the third highest daily temperature recorded since 1985
   - Wildfires burned nearly 10.3 million acres across the U.S. in 2020, the most recorded in 21 years
   - Across agricultural sectors and rural areas, climatic impacts contribute to invasive species and additional costs for weed and pest control, prevented or reduced plantings, decreased health in livestock, losses and damage to buildings, equipment, and land. (USDA, Climate Change Affects)

Given such depressing events, it is nonetheless important to understand that agriculture and husbandry also contribute roughly 20% of the annual increase in anthropogenic emissions of carbon dioxide, methane, and nitrous oxide (i.e., primary greenhouse gases). Various methods are known (but are not as yet globally employed), such as reducing organic material and mineral fertilizers, shifting from traditional to high yielding varieties, switching from rice to some other field crops and, in husbandry, improving nutrition of traditionally managed ruminant animals, improving the management of animal waist, and reducing the burning of biomass. As shown in Table 9.1, greenhouse gas emissions from agriculture and livestock are we significant (Aydinalp & Cresser, 2008). Of course, the actual significance depends on other industries responsible for the other 80% of current anthropogenic emissions.

Table 9.1 Methane emissions from agricultural and other sources in selected Asian countries, 1990 ('000' tons) [5]

Country	Livestock	Other agriculture[a]	Waste	Total	Nitrous oxide
Bangladesh	520	473	74	1335	3
China	8940	18,400	790	33,830	1100
Indonesia	864	2039	—	3746	2769
Japan	520	276	400	1316	54
Kazakhstan	939	—	1763	3555	7
Mongolia	301	—	15	329	0
Nepal	370	542	—	996	1
Philippines	315	559	138	1290	8

[a]Including flooded rice fields

9.2 Emergencies Are Also Opportunities

Every emergency leaves behind its huge measure of human misery, horror, and despair, but each also presents an opportunity to learn something new. Of course, it's up to us to use that opportunity constructively rather than ignore it.

Consider, for example, the Covid pandemic that forced so many of us to stay at home filled with anxiety regarding suspended employment, education of our children, loss of daily, social interactions, and, for many, a chilling loneliness and fear having psychological consequences that may be long-lasting.

Yet, many companies used a relatively new technology (Zoom) that allowed thousands of workers to work at home (many, for over a year) and discovered, somewhat surprisingly, that the productivity of those workers increased. What exactly this phenomenon portends for the future is open to question, but it is very likely that companies will seriously consider significant changes in workplace scheduling to accommodate at-home, virtual work. Just think of some of the consequences: significant reduction in commuting-times; reduction in traffic congestion; lower overhead costs for office space; reduction in costs for child-care!

As a long-time teacher, I certainly realize the limits of current virtual technology in education (especially for young children). However, Zoom is the current technology. Personally, I have no doubt that, especially given corporate interest in virtual work, our current technology can quickly evolve into the 'Star War' standard of hologramlike communication. Such a development may well obviate some of the deficiencies of virtual education.

There are other interesting things the Covid pandemic has revealed by home-confinement. For example, the worst thing one can do in any pandemic is to gather in crowds, especial given the regrettable fact that many people refuse to wear masks. But we still must eat and shop for a multitude of necessities. Restaurants quickly learned to provide take-outs and, soon after, even deliveries. Other entrepreneurs soon appeared to provide limited, specific shopping and delivery services that have rapidly mimiced other very successful companies providing general shopping and rapid delivery. The simple fact is that we have necessarily and rather quickly become used to not driving to a distant vendor or that twentieth century madhouse called a shopping mall to get what we need. All we have to do is press a key on our computer or iPhone and answer the doorbell. Don't 'havta' go out in rain or snow, heatwave or cold, miss appointments or be inconvenienced in any way. Welcome to the twenty-first century!

Well, what does all this have to do with climate change?

Of all the emergencies endured by humans, anthropogenic climate change will (without doubt) prove to be the most globally horrific emergency. We have taken some (if mostly tentative) steps to prevent it and, most importantly, we already have some technologies we can use toward that end, and we are developing others that may prove (depending upon global consensus) to be significant, including:

9.3 Solar Power & Electric Vehicles

Invented in 1952, solar panels designed and constructed in America supplied over 90% of the global market within two decades. Today, America is only one of the world's 10 largest makers of solar cells even though any casual drive through the American countryside will reveal extensive arrays of solar panels that signify an early awareness of the necessity to move away from fossil fuels to solar power. However, political decisions in the 1970–1980s related to a foundering economy made it increasingly difficult for America to compete competitively with Japan and, subsequently, China in the global solar market. The result has been that American concentration on solar power is today primarily in research and development (R&D), the province of scientists. It has been suggested that any dramatic resurgence of the solar industry in the U.S. requires a melding of both scientists and engineers (Meyer, 2021). Personally, I see this suggestion as an acclamation of the historic success of NASA's placing humans on the moon within the 10-year limit promised by President Kennedy…a feat made possible only by political and economic support of scientists and engineers working together toward a common objective.

Of course, besides the mere scientific and engineering challenge of further development of solar power is the economic advantage. The fact is that, as I write this paragraph, the average cost of electricity generated by fossil fuel is between 5 and 17 cents per kilowatt hour (KWh), while that generated by solar power is between 6 and 8 cents per KWh. This is a good example of how a technology motivated by one concern (i.e., cost of fossil fuel) proves to be a major tool for mitigating another concern (namely climate change). Another example is the electric car.

Small-scale electric cars first appeared in the first quarter of the nineteenth century, when land-transport was by horse and buggy. It took almost 75 years for the first electrified carriage to appear. Within a decade, electric vehicles became rather common, due especially to Thomas Edison's improvements of batteries. However, the discovery of cheap crude oil quickly ended their popularity. By 1930, electric cars were replaced gas-driven vehicles until, in the 1960s and 70s cheap gas disappeared and serious interest in electric vehicles was reestablished in frustrated drivers thoroughly tired of high prices, rationed gas, and long hours trapped in gas-station wait lines. Key environmental legislation in the 1990s also prompted increasingly negative attitudes of the American public regarding fossil-fuel-contamination of the environment. Since the turn of the century through the present, the technology of electric vehicles has steadily evolved in both design and application (Matulka, 2014). Again, I repeat: the experience of one type of emergency (i.e., loss-of and skyrocketing cost-of gasoline) can provide a means of ameliorating another type of emergency (e.g., climate change).

9.4 Agriculture & Husbandry

At the beginning of this chapter, I stated that 14% of anthropogenic contributions of the greenhouse-gas Methane was due to agriculture and livestock. However, it is also true that regional agricultural plants and livestock will be dramatically affected by that same climate change. An appropriate simile is the owner of a business who, on vacation, lights a campfire, sparks from which escape to ignite a wildfire which, in turn, burns down his place of business. Well, it happens!

Consider another issue related to agriculture. In 1972, the world population was about 3.8 billion; in 2022, 8 billion; and projected for 2072, 10.5 billion. These 50-year intervals clearly show the desperate need for more and more food, a persistent need which is evident over even earlier centuries.

Given the magnitude of global population-growth, it became obvious by the late 1960_s that food-shortage would kill hundreds of millions in the 1970_s and '80_s. This realization resulted in what is known as the "Green Revolution" which, through ***selective breeding***, dramatically increased the productivity of grains such as rice and wheat, doubling yields over a thirty-year period. Today, we use a new technology (***GM*** for ***genetic modification***) based on major advances in molecular genetics to further increase crop-yield. In the United States, about 90% of our corn, cotton and soybeans are genetically modified (Folger, 2021).

A good example of how GM can be used to ameliorate crop loss due to climate change is the tomato. It grows best between 64 °F and 77 °F. Given climate change, about 66% of land in California (a major source of tomatoes) may not be suitable due to climate change. Using a molecular tool (**Crispr**)[5] to modify the heat-tolerance of tomatoes (and, coming soon) of many other crops certainly marks the beginning of a second food-revolution (Zaraska, 2021).

Another rather recent technology regarding agriculture is Vertical Farming (FM) which was developed to maximize food production in a world where traditional farming is limited by available land. However, there are other advantages that

[5] Crispr is the acronym for 'Clustered Regularly Interspaced Short Palindromic Repeats', a molecular procedure for editing genes invented in 2012.

continue to expand its utility. FM involves a self-contained controlled environment (CE) that can be multi-tiered in a tall building and equipped for hydroponic (i.e., soil-less; therefore, water based) growth. This is a constantly developing approach to farming, with ongoing studies regarding types and quality of crop, energy efficiency, and economics (Beacham et al., 2019).

With climate change, various traditional regions of major crop-production will become unusable and alternative locations unavailable. Vertical farming, however, could be used in a variety of environments, including cities or localities not suitable for traditional farming. Cities would appear to be ideal locations because of the huge cost-decrease associated with cross-country transport of produce and the greatly enhanced accessibility by consumers. In fact, such advantages alone provide sufficient incentive for such a development even in the absence of climate change. Well, that depends on the variety and quality of crops and VF technologies regarding such issues as air quality, water recycling, and carbon-footprint.

With respect to husbandry, new technologies seem to appear at breathtaking pace, especially with regard to bioengineering (e.g., Crispr); another recently invented tool of bioengineering is **stem-cell technology** which, over the past few years, has spawned a whole new industry best described as lab-grown meat-production.

Stem-cells are cells that produce other highly specialized cells (e.g., brain, muscle, bone, blood, etc.).

We know how to coax stem-cells derived, say, from a cow into cow-muscle. In simple terms, we don't have to slaughter a cow to make a cow-burger; we can grow that burger in a laboratory.

The first burger was lab-grown in 2013; since then, various companies have since improved the taste and density; A Dutch company has developed the process to allow muscle and fat cells grow together, thereby improving taste and texture. It has been predicted that 35% of all meat consumed globally will be cell-based by 2040 (Davies, E., BBC ScienceFocus).

Just think of the literally millions of cows, chickens, and pigs that are raised for our tables per year – the amount of greenhouse gases expelled to the atmosphere by their gaseous, liquid and solid wastes, by the myriad food-processing technologies employed, and by the transport of the final product to market.

In the United States, the carbon-footprint using traditional technologies for providing meat, vegetables, and grain to our tables in 2019 was at the least 10% of America's production of greenhouse gases (EPA, 2021a, b).

9.5 Summation

While the COVID pandemic has been (and continues to be) a fountainhead of social and political foment, economic disarray, and personal agony, climate change will be worse; it will also last for a much longer time. If decisions are not now made to significantly reduce the human contributions of greenhouse gases to the atmosphere, earth's temperature will continue to rise to a tripping point beyond which we will have no control.

The major problem is that such decisions must be agreed upon by all nations. Clearly, obtaining global consensus is always difficult (to say the least) and demands Herculean effort. There have been periodic efforts in this regard, and, over the past few decades, some progress has been (and continues being) made. Any real reduction of humanity's carbon-footprints depends first and foremost upon the decisions made by individual nations. But nations do not make decisions; it is those humans who, in complex arrays of governmental, public, and private institutions and, at the core, individuals and their medley of congeries who ultimately make their own decisions that may or may not influence others.

Suggested Exercises for This Chapter

Select from among the following topics and, using the Internet, research topic and prepare either an expository oral presentation or written essay:

- Famine in India over at least the past 5 years
- Frequency and magnitude of hurricanes in southeast United States
- Heat waves in past 3 years in western United States
- Glacial melt in Arctic
- Changes in the Gulf Stream
- United States participation in international efforts regarding climate change
- Likely effects of climate change on human health
- Effects of climate change on wildfires in the Arctic

References

Aydinalp, C., & Cresser, M. (2008). The effects of global change on agriculture. *American-Eurasian Journal of Agricultural & Environmental Sciences, 3*(5), 672–676.

Beacham, A. M., et al. (2019). Vertical farming: A summary of approaches to growing skywards. *The Journal of Horticultural Science and Biotechnology, 94*, 277–283.

Climate Central. (2021). Fourth national climate assessment: Chapter 14. In *Human health, climate change and health.*

EPA. (2020). *The legacy of EPA's acid rain research.*

EPA. (2021a). *Climate change indicators.*

EPA. (2021b). *Sources of greenhouse gas emissions.*

Folger, T. (2021). *The next green revolution.* National Geographic.

IPCC. (2021). *Intergovernmental Panel on Climate Change, IPCC Press Release: Climate change widespread, rapid, and intensifying.*

Lindsey, R. (2020–2021). *Climate change: Global sea level.* Climate.gov. science & information for a climate-smart nation.

Metulka, R. (2014). *The history of the electric car.* U.S. Department of Energy.

Meyer, R. (2021). Why America doesn't really make solar panels anymore. *The Atlantic.*

National Center for Environmental Health. (2020). *Diseases carried by vectors.*

Pescaroli, G., & Alexander, D. (2015). A definition of cascading disaster and cascading effects: Going beyond the "toppling dominos" metaphor. *Global Risk Forum (GRF), Planet@Risk, 3*(1).

Rohr, J., et al. (2011). *Frontiers in climate change-disease research.* Trends in Ecology &Evolution-Cell Press.

Schuur, E. A. G., et al. (2013). *Expert assessment of vulnerability of permafrost carbon to climate change*. Springer.

Science News. (2022a, February 4). *Preventing pandemics costs far less than controlling them.* Duke University.

Science News. (2022b). *Climate change has likely begun to suffocate the word's fisheries.* American Geophysical Union.

The Guardian. (2011). *What are the main man-made greenhouse gases.*

The Guardian. (2021). *Scientists watch giant 'doomsday' glacier in Antarctica with concern.*

University of Leeds. (2021). Himalayan glaciers melting at 'exceptional rate'. *Science News.*

World Meteorological Organization. (2021). *WMO recognizes a new Arctic temperature record.*

York University. (2021, October 21). Northern lakes warming six times faster in the past 25 years. *ScienceDaily.*

Zaraska, M. (2021). *The tomatoes at the forefront of a food revolution.* British Broadcasting Corporation.

Part III
Social-Structural Considerations

1.1 Preface to Part III

The following chapters deal entirely on decision-making in both emergency planning and response. Having previously focused on practical issues inherent in a holistic approach to the prevention and amelioration of emergencies, my intent in Part III is to focus on difficulties regarding relevant governmental and other institutional decision-making. In no way do I intend to censure, impugn, or otherwise berate decisions made in the past; rather, by discussing those difficulties, to prod perhaps new thinking and especially new resolve.

Chapter 10
On the Scope of Emergency Planning

10.1 Prologue

Emergency Planning obviously precedes Emergency Response. This would appear to be a simple statement of fact. However, it is too general and misleading a comment because, in the United States, for example, planning for an emergency is conducted at a myriad of administrative levels, including National, State, County, and Municipal levels. By any measure, that's a lot of decision-making that has to go on to define the scope and duration of any specific emergency response. An appropriate question is: "How are decisions by so many different authorities integrated to achieve first, a coordinated response; second, an effective and comprehensive response; third, a response that is also commensurate with purported societal values and ideals?"

There are 195 countries in the world, each having its distinct cultural heritage and political and social structure, each, its own way of doing things and for its own reasons. Yet, … **YET**, despite the vast variety of human societies, they all experience emergencies of one kind or another, be they local or global. That is a significant commonality among us all and has been throughout our many thousands of years of human ancestry. Our response to any emergency, whatever and wherever that emergency may be, is a significant *measure of our human capacity to care for others, many times, regardless of personal risk.* Ah, therein lies a persistent problem!

Of all the species of animals we know of, the human animal, with its large, cognitive brain, is unique (as best we know) in its ability to think and act in extremely diverse directions, with a seeming infinity of contradictory purpose and motivation. For the many thousands of years that our earliest ancestors lived their lives in groups of three-plus dozens, daily survival of such physically puny creatures depended upon mutual care, regardless of individual desires and personal motivations; today, we increasingly live amid thousands of strangers, all of us being daily dependent mostly on easily available technologies rather than communal interactions.

P. A. Erickson, *Effective Environmental Emergency Responses*,
https://doi.org/10.1007/978-3-031-05893-6_10

In previous chapters, I have given particular attention to the fact that the duration of an emergency is not to be determined by local or even national newscasts, but by the duration of its affects. I repeat…the fire may be over, yet, what of the many firefighters who are uninsured and not medically monitored at least yearly for cumulative lung or other organ damage caused by smoke or exposure to toxic fumes? What of victims (including first-responders) who undergo trauma with prolonged (if not permanent) psychological affects that last well beyond the short span of public interest? The flood or hurricane or epidemic is over. But what of those who disproportionately suffer the consequences of an emergency (e.g., pandemic, flood, etc.) primarily because of social and institutional inequities?

We are finishing the first quarter of the twenty-first Century. I offer, to you who will live your lives though the rest of this century, the following quote as a reminder that we have a lot of work ahead of us if we want effective and comprehensive emergency response.

> The worst sin toward our fellow creatures is not to hate them, but to be indifferent to them: that's the essence of inhumanity (George Bernard Shaw).

With that, I turn now to various factors that exert manifold influences on decision-making in the process of emergency planning as well as the execution of the product of those deliberations.

The Socialization Process Otherwise known as 'Enculturation', socialization begins at birth and persists over the various timeframes of prolonged development. A key aspect of socialization which is relevant to human decision-making has been explained by various natural and social sciences and may be succinctly summarized.

Within 5 years of their birth, human infants have learned the language and customs of any society to which they're borne. In consequence, they have absorbed the values and prejudices that will stubbornly influence their latter development. Having absorbed this information at such a young age, they will tend to forget how they learned how to think and act as they do, and simply accept their attitudes and assumptions as they do their arms and legs (Tattersall & Desalle, 2019).

The point here is that many of us are quite comfortable with what we have learned up through our adolescence and perhaps early adulthood, and essentially stop considering new ideas and values. Meanwhile, we all grow older, assume various levels of social responsibility and authority, and make decisions that are influenced unknowingly by our own distant-past and now forgotten experience.

Socialization is a necessity Nature has devised to supplement the very few muscular instincts with which we are born. After all, we are born knowing only how to suckle and grab. We must learn everything else. The early childhood learning we all experience is the most important for ensuring survival of the otherwise helpless babe. Unfortunately, what a child learns can become 'The Truth' imbedded in an infant brain which will not be fully developed for almost another 20 years…… a period long enough for the brain to turn fairytale, disguised as truth, into prejudice.

Vested Interests This phrase refers to a wide variety of 'personal stakes' we protect, including physical, psychological, and social investments we strive to maintain or enhance such as wealth, reputation, power, and whatever else serves our sense-of-self. We all have vested interests to greater or lesser degrees. They are the self-maintenance impulse of any cognitive brain and, thereby, essential factors in personal and social decision-making. I offer the following as a example of their influence in emergency planning.

Several years ago, a colleague[1] and I presented a 40-h training session for a group of emergency-response professionals in Southeast Asia. They were employees of half-a-dozen major petroleum companies who controlled the transport of liquified natural gas (LNG) through a port-city having a population of 5.1 million.

LNG is extremely dangerous; container ships that transport LNG are rightfully called floating bombs…an explosion on a single ship can destroy a city! We, therefor, had asked our students to provide us with their written plans for dealing with a fire-emergency on the city's wharf and were both horrified by the fact that each of their plans described contractual allowance with other LNG transport-companies along the same wharf for borrowing necessary firefighting and containment equipment from each other.

In such a fire in the vicinity of ships loaded with LNG, no company in the same business and location is going to lend out its essential equipment. Rather, contract or no contract, it will use its own equipment to look after its own immediate threat and, in a situation involving LNG, focus its efforts to bulldozing nearby and threating enflamed structures into the sea and getting its own endangered ships out of port-waters as fast as possible.

What does this have to do with vested interest? Our students, who did not have executive power (but a surplus of responsibility), had a vested interest in keeping their jobs; likewise, so did their immediate and successively higher bosses who, in turn, had to satisfy still higher company executives located half-way across the world and having primary responsibility for monitoring corporate finances.

As I said when discussing plate-tectonics (Chap. 8), pressure has consequences. Well, so does our avarice regarding money and employment-security, two persistent factors in hierarchical decision-making in so many ways, including emergency planning and response.

Another personal experience comes to mind regarding the nature of hierarchical decision-making.

My older brother was my life-long best friend and confidant. In my earliest memories of him, I knew that he was resolutely intent on becoming a physician. He did become one, a general practitioner which, in those days, was the "family doctor" who made house-calls whatever the time of day or night and, while providing his care, never looked at the clock. But within a decade or two (1960s–1970s), medical

[1] Datuk Dr. Soh Chai Hock: National Chairman of the Malaysian Fire Prevention Council; Director General (Rtd) of Malaysian Fire and Rescue Services after 36 years' service.

science became so much more complex and highly specialized, the very definition of "general practitioner" changed (CDC, 2000).

From being an independent and largely self-sufficient provider of healthcare, the general practitioner was transformed into an essential turnstile in an intricately complex healthcare system characterized by organ-specialization, advanced surgical procedures, and molecular technologies and, of course, a matrix of insurance providers and other overseers skilled in the practice of Law.

Because of these changes in the medical system, my brother felt he was spending as much (if not more) time filling out insurance forms for his patients and meeting other legal requirements than he did providing his patients with medical care. Frustrated and disheartened, he retired several years before he might have due simply to age.

His world had been abruptly changed by the phenomenal increase in medical knowledge and technology and their associated vested interests. And so has it happened in many other professions and types of work that are suddenly transformed by new knowledge and subsequent technologies, including for example, manual home-construction now transformed by 3-D printing, or car-repair by in-vehicle and in-shop computers. With each transformation comes a coterie of vested interests that promulgate that new knowledge and technology No, it's not all about money, though money is obviously a constant factor. Vested interests also include self-identity (as with my brother), social status, power, and the plethora of other intangibles conjured up by the human brain.

Major advances in the sciences and subsequent technologies have steadily accelerated from the mid 1950s through the present. Consider how many times in the past 5–8 years you have had to replace your I-phone or your home computer. How often within a month do those internet-spawned devices automatically receive updates?

When did it become possible for two people with amputated arms shake robotic hands simply by thinking about shaking hands? 2017. When did hobby-drones become available to the general-public? 2019. When was the pig-heart successfully transplanted in a human? 2022. And so on and on!

Our world is changing more and more rapidly because our knowledge is steadily increasing. But does new knowledge and subsequent new technologies necessarily result in new thinking?

Habit and Tradition A late seventeenth century concoction, the English word routine is derived from the French '*route*' (road) and the ancient Greek and Roman suffix '*-ine*' (of, or pertaining to') thusly interpreted as, 'the way or road usually taken' or, in the more common vernacular, a habit, as in habitual behavior. While we most often speak of habits of particular individuals, (e.g., smoking, eating peas with a knife, taking a 1-mile walk in the morning before work, etc.), individual habits are also shared by many others.

While some individual habits develop spontaneously or by personal experience, many habits are taught, such as brushing teeth before going to bed, using a bookmark rather then folding page-corners, and writing thank-you-notes for every present received. It therefore follows that some habits have to be learned.

Of course, learning is not the sole province of formal education; personal experience outside of schools is sometimes most instructive.

For example, Benjamin Franklin had only 1-year in grammar school and 1-year private tutoring. In spite this paucity of formal education, he attained National and International acclaim as a writer, printer, scientist, inventor, colonial and international ambassador, the prime founder of the University of Pennsylvania, and an essential member among the Founding Fathers of the United States.

How did he accomplish so much? As a young teenager and throughout his life, he developed the habit of searching out those whom he considered ingenious and wise, and emulating their diverse ways of thinking, always added his own spark of inherent ingenuity.[2] He was born in the beginning of the eighteenth Century (1706) and died almost at its end (1790). Chronologically, a child of the eighteenth Century, intellectually, he not only prepared his fellow colonists for the nineteenth but acted throughout his adulthood to ensure that the turn of centuries for Americans would signal a completely new and better way of life. If only the habit of eagerly seeking out and learning new ways of thinking rather than blindly perpetuating the past were to become so common as to become a tradition! After all, traditions are basically long-term habits that we typically accept without question, even in the face of sustained assault. They provide a sense of solidarity to social groups, influence human character, and conduct and, if ignored, can obliterate the sense of purpose (Frohnen, 2001).

Whether National or cultural, State, or familial, religious, or secular, traditions can promulgate behaviors over a wide range of acceptability by others, as exemplified in the extreme by the traditions 'honor killing' and 'Thanksgiving'. Like all habits, traditions are subject to change, but that requires a change in thinking followed by a steady resolve to change behavior.

A good example of some of the pitfalls associated with tradition-change is the evolution of colonial allegiance to England. Prior to any thought of revolution, colonists (whatever their ancestry) viewed themselves as citizens of England and therefore enjoyed the rights of English men and women whether living in England or in an English colony. However, King George and the English parliament implemented a series of Acts and policies that required the colonies to pay certain taxes and otherwise depressed American trade. Not having any representation in Parliament, the colonists did their best to convince the English government to cease its odious treatment of its colony, arguing their right, as English citizens, "No Taxation Without Representation", but to no avail. In fact, England responded with military action that resulted in the battles of Lexington and Concord in the spring of 1775, lit the short-fuse of rebellion, and divided American colonists into two camps – those who retained their allegiance to England (the 'loyalists') and the rebels (the 'revolutionaries') who renounced their traditional allegiance and opted for war against England. But first came a civil war waged by the revolutionaries against the loyalists,

[2] I highly recommend you read: "Young Benjamin Franklin: The Birth *of* Ingenuity", Nick Bunker, 2018, Borzoi Books, Knopf.

designated traitors to America and treated accordingly. Shunned, threatened, attacked, incarcerated, and murdered, tens of thousands of survivors fled mainly to Canada and England, but at least several hundred thousand remained in America (Ellis, 2021).

Yes, changes in tradition can result in both positive and negative social consequences; very often, however, they are unpredictable.

Mass & Social Media Mass Media date back to colonial times in the United States with newspapers and almanacs that passed information on to the general public. Over generations, printed news was supplanted with radio, television, and cinematic technologies. Social Media (e.g., iPhone) are offspring of current digital technologies whereby the general public is both the creator and receiver of information.

Given their early creation, Mass media have co-evolved with the development of various legal constraints, professional standards, and cultural expectations. Social Media are currently undergoing increasing scrutiny regarding possible regulatory control. However, societal concerns regarding both mass and social media (mass-social media) are prompted by a variety of issues, including the frequency of transmissions of misinformation and outright lies, personal bullying, racial, ethnic, and gender discrimination, and political bias.

Mass media have great influence on individuals, institutions, and society. But who is it that decides how media wield that influence? Who owns the media? Who has control.? Who is questioning that control (McQuail, 1979)?

Various parties have strong incentives to exploit media in furtherance of private goals. How do the goals of governments, politicians, advertisers, and other special interest groups translate into deliberately biased information (Enikolopov & Petrova, 2017)?

Some American teens consider social-media 'extremely' or 'very' important for them for expressing themselves creatively; some have reported being cyberbullied...... others report being subjected to violent images and pornography, or to racist, sexist, homophobic, or religious-based hate content in social media (Summer, 2019).

In short, we make decisions based on what we experience, on what we know and/or what we think we know, depending on what we have absorbed into some matrix of our formal education and our personal and private, experiential learning. Either way, whatever we have learned is kneaded, molded, and sculpted by our socialization, vested interests, personal habits and handed-down traditions, and the omnipresent mass-social media. We often mistake a stranger for a relative or friend, they look so alike. Ignore the body; if we could somehow envision that person's dynamic brain humming away with its 100 billion neurons, each of which has up to 15,000 connections with others, we would immediately realize that every human is unique. The body is the brain's tool; the brain is the craftsman – that's where the real person is.

Figure 10.1 summarizes the uniqueness of each individual human as influenced from birth to adulthood by the factors discussed above. Note that I added the genetic contribution of 20,000–25,000 genes (half from each parent), which is within the

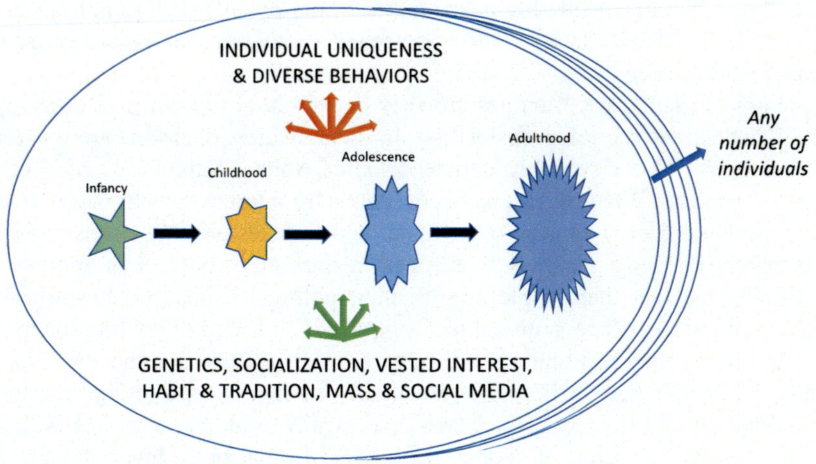

Fig. 10.1 The socialization process

range for a grape and chicken, but which is quite sufficient to establish the biological uniqueness of every baby. I've also indicated that what is true for one individual human is true for all, … thus the diversity of behavior so evident in the arduous task before any group of humans attempting to make consensual decisions.

While our long-ago ancestors lived in groups of three-plus dozen, consensual decisions were facilitated by a paucity of technology, few alternative objectives regarding where and when to find sustenance and safe-haven and, their small numbers. The choice for making the right decision was whoever was obviously the most successful forager and/or scavenger or hunter. Someone else might be the obvious choice for finding safe-haven. There was no one person always in charge of everything; day-to-day leadership was accorded to the individual who was the most successful in providing what was immediately needed for survival (Gray, 2011).

Such egalitarian societies prevailed for over 90% of human ancestry; now there are very few (a bit over one-tenth of 1% of the world's total human population), with even those tiny few dwindling toward extinction. Today, the rest of us live in dense population-centers with our seemingly countless technologies and organized by diverse institutions having variously defined degrees of authority.

10.2 Top-Down & Down-Up

These terms are generally used to denote thinking from the general (top) to the specific (down), and the reverse. In that sense, there are various applications. For example, in business, the owner or other highest authority (i.e., 'top') makes a decision that becomes a rule governing employee-behavior (i.e., 'down'), and in investment-related enterprises, investing focuses on macro factors of the National economy (i.e., 'top') before considering micro factors (i.e., 'down') such as individual companies. In these two examples, decision-making may proceed in the reverse direction (i.e., down-up).

Whole societies are often described as being, either top-down or down-up, but such distinctions are often misleading. In dictatorships, decisions are made at the top authority who is variably christened king, Sultan, Supreme-Leader, or more simply, Dictator. Whatever the name, decisions from top to down (otherwise known as common citizens). On the other hand, democracies are often considered to be essentially down-up societies. Well, true in one sense, but not in another. True, regarding the fact that people vote in their political leaders (a down-up event). However, once in office, political leaders make laws for all of us (top-down).

Americans may institute a recall of certain governmental authorities, but some (e.g., Supreme Court Justices) are essentially exempt except through the tortuous procedure of impeachment which was successfully used only once in U.S. history.[3]

In the private sector, of course, there is no formal procedure available for an employee to fire her/his boss or a member of a Board of Trustees. As with formal recall and impeachment, the dismissal of high authorities demands massive and persistent public outcry. Even that does not guaranty success, but at least it can serve as a clear signal that change is needed and publicly supported.

In our daily lives we make many decisions without concern or required consultation but, when dealing with societal institutions (e.g., governmental, legal, financial, educational, religious, etc.), we quickly learn to subvert personal inclinations to established protocols. This is the essence of social organization … that we learn to follow the rules others have decided to make. However, as necessary as such rules are, it is also necessary to at least question their propriety. After all, there was a time in the United States when the Supreme Court established "Separate but Equal" as a sop to racism; when woman could not vote; when corporal punishment was a basic tool in early education; when same-gender sex was a crime.

To question behavioral-rules laid down by others has long-been the American and, I presume to say, an inherent human propensity as demonstrated by the fact that one of the first words a toddler learns to use is "why?". How can we learn anything if we don't ask 'why'?

10.3 So …, What's an Emergency?

I began Chap. 1 of this volume with a preliminary focus on the meaning of 'emergency', an English word adopted in the mid 1630s from the Latin 'emergere' (*to rise out or up*). Whether unforeseen or predicted, an emergency is historically defined as an *occurrence requiring immediate attention*. In the modern vulgate used by the mass-media and, thus, the general public, an emergency is often simply defined by the name given to the 'occurrence', as in hurricane, flood, pandemic,

[3]Associate Justice Samuel Chase in 1805 on charges of drunkenness and insanity.

wildfire, etc. Personally, I consider the vulgate unfortunate. Whether the 'occurrence' is a phenomenon of Nature (e.g., hurricane) or of Human origin (e.g., mass-shooting), or even of both (e.g., wildfire), the focus has to be an immediate attention to the totality of effects of the 'occurrence'. Please note that 'immediacy' connotes 'urgency'; neither implies *duration of that attention*…only the start of it. I also refer you back to Chap. 2, specifically my discussion of 'vulnerability' and Fig. 2.2 which focuses on key elements of risk assessment expressed in the formula HAZARD + EXPOSURE = RISK.

Hazard is the *possible* harm or injury (associated with 'occurrence') and **Risk** is the *probability* of humans experiencing that harm or injury. It is **Exposure** that determines the transformation of the possible into the probable, and exposure is measured over days, weeks, decades and even (in terms of genetic damage) over generations. Why then are emergencies said to be over in the space of a few days or weeks? Why are emergencies compared by the costs of restoring structures and/or the costs incurred by insurance companies and, typically, not the long-term costs of treatment and rehabilitation of victims (including first responders)? Why do cities continue to expand into earthquake- and tsunami-prone areas? Why do we wait for an 'occurrence' to happen before we respond to subsequent emergencies when we might possibly implement preemptive preventive actions?

So many whys! How to answer? I suggest that the wrong response to any of them is the proverbial "Well, that's what we've always done".

Suggested Exercise for Chapter 10

Using the Internet or printed sources, select any emergency (within last 6–8 years that received National and/or international attention and was the subject of extensive news coverage. Prepare a presentation (oral or written) that focuses on those response efforts that received (A) significant news coverage and (B) little or no news coverage with respect to the following issues:

1. Estimated cost of replacement and repair of structures
2. Estimated cost of first aid provided to victims (including responders) and cost of likely prolonged health issues
3. Inequitable distribution of property- and human-losses regarding (A) color or ethnicity and (B) income levels

Given the information you have gathered, include in your report your personal reactions and conclusions.

References

CDC. (2000). General practice—Time for a new definition. *The BMJ*.

Ellis, J. J. (2021). *The cause: The American revolution and it's discontents, 1773–1783*. W. W. Norton & Company. (Original source: An Address to the People of Great Britain, American Archives).

Enikolopov, R., & Petrova, M. (2017). *Mass media and its influence on behaviour*. CREi (Centre de Recerca en Economia Internacional).

Frohnen, B. (2001). Tradition, habit, and social interaction: A response to Mark Bevir. *Humanitas, 14*(1), 108–116.

Gray, P. (2011). *How hunter-gatherers maintained their egalitarian ways*. Psychology Today.

McQuail, D. (1979). The influence and effects of mass media, mass communication and society. .

Summer, A. (2019). *Social media's growing impact on our lives*. American Psychological Association.

Tattersall, I., & Desalle, R. (2019). *The accidental Homo Sapiens*. Pegasus Books.

Chapter 11
Consensus vs. Conflict

Taken in 1790, the first U.S. census yielded a population of 3.9 million; by 2022, it was 300+ million, and the current projection for 2058 is 400+ million. Estimated world populations over the same period are ≈1 billion (1800) and ≈10 billion (2060). While there is a plethora of ramifications of the proliferation of humans, a species comprised of unique (and often argumentative if not downright cantankerous) individuals, one aspect of particular importance is how heterogeneous groups can come to consensus on any issue but, especially on the scope of emergency planning and response.

By way of example, I return to the Influenza pandemic of 1918–1919 (Chap. 5) and the current Covid pandemic.

The Influenza pandemic occurred before we had any scientific knowledge of viruses, even though we were aware that there existed some 'disease-causing-particle' which was substantially tinier than bacteria. What that particle was, nobody knew. Our understanding of viruses did not begin until the invention of the electron-microscope after World War II, over a quarter of a century after 30–50 million deaths world-wide and about 700+ thousand in the United States.

Vaccines were produced, as were face masks, and death rates in some cities decreased, at least for a while, but the vaccines soon proved to be immunologically ineffective, and face-mask-ordinances were passed. At first, they seemed to be effective. However, they were made of gauze materials that could not provide a small enough mesh to prevent the penetration of a sub-microscopic viruses. Nonetheless, as the saying goes, "Better than nothing". For a while, at least. But, only for a while. People began to complain that masks were humiliating, they fogged glasses, and, moreover, mask-mandates were declared to be unconstitutional. Even the threat of arrest (30 days in jail) or fines ($5) did not dissuade growing numbers of citizens who refused to wear masks or wore them under their chins despite the warnings of prominent health specialists. Ninety percent of San Franciscans ignored the mayor's plea to readopt the wearing of masks; various

P. A. Erickson, *Effective Environmental Emergency Responses*,
https://doi.org/10.1007/978-3-031-05893-6_11

groups congregated into an 'Anti Mask League'. Public resistance to wearing masks and threats of violence forced the San Francisco Board of Supervisors to hold a public hearing with hundreds of citizens; after several days of mostly argument, the Board voted against a masking ordinance (Crosby, 2003).

Now, 104 years after that Influenza Pandemic, we have the Covid Pandemic. What have we learned in the meantime that will serve our survival in future pandemics that are certain to continually threaten humanity?

Our sciences are pushing forward the frontiers of biological knowledge at an increasingly accelerated pace, all the while generating new knowledge and, thereby driving the development of previously unimagined, highly sophisticated, and effective technologies. Modern bacteriology, virology, epidemiology, and psychology are at the forefront of the Medical and healthcare practices that include care of both the physical and the psychological sphere of human life.

However, while medical science has undergone phenomenal advancements, human behavior regarding pandemics has not changed; in fact, driven by the proliferation of mass-social media, as well as by the long-standing, public aversion to the intricacies of science, the response of the American public to Covid has been disheartening (to say the least), and (more to the point) contra-productive to timely and effective pandemic-response.

During the 1918–1919 pandemic, the means of public communication included radios, telephones[1] (requiring so many operators who became sick that service was severely reduced), newsreels (shown first in music halls; subsequently in movie theaters), telegraphs, and even signal lamps. At best, news travelled slowly, was unavailable or too expensive for many, or otherwise intermittent.

I need not compare in detail the effects of communication in early twentieth Century with those of today, except by penning the adjectives *global, instantaneous, reliable, readily available* which, if considered together, expose a simple fact of the twenty-first Century – that anyone can spread the truth, or misinformation and outright lies to anyone anywhere at any time. It is important, also to realize that the U.S. population in 1918–1919 was roughly 103 million; presently, it is 330+ million – more than a three-fold increase in the number of Americans who have direct access to the most modern communication systems.

It is evident that previous public disavowals of appropriate means of reducing rates of infection by the Influenza-virus (e.g., masking and social-distancing) have not only resurfaced but also become more pervasive and determined in our current Covid pandemic. This is not an accident but, rather, the direct consequence of a much larger population of unique individuals with easy access to the most efficient communication systems ever invented.

I have no doubt that Covid will be long remembered not only as a debilitating and deadly disease but also as an *opportune excuse for offensive, aggressive, illegal, or otherwise socially disruptive behavior* such as:

---

[1]Note: In 1920, telephones were available in only 30% of American homes.

- Verbal and physical attacks by mask-less citizens on the masked, including shaming, bodily injury, even murder
- Physical and social abuse of citizens whose national origin is thought to the source of the pandemic
- Politicization of health-related issues (e.g., masking, social distancing, education-related decisions, etc.), including mayors countermanding governors and, in turn, governors countermanding national authorities
- Mass-social media publicly demeaning acclaimed health-scientists and, and simultaneously promoting the purported expertise of those who possess little or none that is directly relevant to a pandemic

What's the lasting American legacy of Covid? Two footprints in the sands of time: one deeply engraves forever the deaths of over 900,000[2] and the varied illnesses of many thousands more; the other shows the *toes* of confusion, the *arch* of anger, and the *heal* of social-disfunction, pointing in the direction away from social solidarity toward anomie. I'm reminded of a sardonic epithet widely shared among experts in infectious diseases who have studied the American response to the Influenza pandemic –

> The nineteenth century was followed by the twentieth century, which was followed by the … nineteenth century. (Crosby, 2003)

Given our response to Covid, we might as well say that the twenty first century is likely to be followed by the twentieth.

Well, maybe. There's always hope. In fact, just 2 days ago (as I write these words), a group of epidemiologists, economists, ecologists, and biologists representing 21 institutions published their finding that we can reduce the risks of future pandemics by investing as little as 1/20th of the losses thus far incurred by Covid in conservation measures designed to stem the spread of zoonotic viruses from wildlife to humans. Such measures include ending tropical deforestation and the worldwide trafficking of wildlife, cessation of wild-meat trade, and the improvement of global disease-surveillance and control of both wild and domestic animals (Science News, 2022a, b).

The primary motivation behind these recommendations is the fact that epidemics (i.e., localized pockets of disease) are now occurring more frequently and, given the volume of international travel, spreading more easily from continent to continent, thusly becoming a global emergency; after all, it only takes a single flight to initiate the transformation of an epidemic into a pandemic.

Of course, it's one thing to recognize or invent a way to prevent or ameliorate an emergency; it's quite another to achieve a sufficient consensus to do it.

[2] As of February 5, 2022, 2 years after the first confirmed American victim of Covid 19, the U.S. death-toll exceeded 900,000; 13 months into the United States' vaccination campaign only 64.3% of the U.S. population has been fully vaccinated.

11.1 Modernity: An Institutional Maze

In Chap. 10, I began a discussion of institutions which I want to expand upon regarding societal decision-making, especially with reference (in several other chapters) to our early ancestors who lived their lives in small groups dictated by the carrying-capacity[3] of their environment, and their rather casual, situational approach to authoritative decision making. As with us, they made many personal, day-to-day decisions. We, however, are also deluged with decisions made by others who are elected or appointed to, or otherwise assume leadership in a wide variety of institutions that are tightly intertwined in that hierarchical complexity we denote as a 'social system' which exerts not only powerful influence on our behavior, but also entails various degrees of monitoring and even punishment of that behavior.

It should not be a surprise that a significant number of Americans suffer depression, bipolar disorders, anxiety and even fear. Whether a workplace requirement, a governmental edict, a school-decision, a National or municipal vote, an investment calamity or any other institutionally created, promoted, or sanctioned mandate, each creates a current that races down the human-synaptic skein in search of manifold open-gates to a potpourri of pre-primed emotions.

Decision-making in a group of 30+ (our ancestral lineage) was situational; leadership depended on recognition of an immediate survival-need and on whatever individual was known to have the most pertinent skill to satisfy that need; their tomorrow would reveal a different need and, thus, the skill of someone else. Leadership was not a profession; it was *a specific task to be performed at a specific time*.

However, hierarchal decision-making in groupings of millions, whether dictatorships or democracies, requires leadership to become *a profession with attendant personal qualities of character* – a necessary tool for fashioning consensus out of the cacophony of conflicting personalities, personal experiences, desires, and prejudices.

Key factors directly influencing this historic transformation of the concept of 'leader' include variations in dominance (of individuals and/or groups) within large populations, personality, experience, and various constraints imposed by the time it takes to gain consensus within a large population comprised not only of unique individuals, but also of their diverse social networks (Gavrilets et al., 2016).

The following quotes epitomize ideal character-traits commonly ascribed to a modern leader:

- *John Quincy Adams*: "If your actions inspire others to dream more, learn more, do more and become more, you are a leader."
- *Ralph Waldo Emerson*: "Do not follow where the path may lead. Go instead where there is no path and leave a trail."

[3]Carrying-capacity: the amount of food, water, and shelter provided per acre of environment to support a certain number of a particular species.

- *John F. Kennedy*: "Leadership and learning are indispensable to each other."
- *Woodrow Wilson:* "You are not here merely to make a living. You are here in order to enable the world to live more amply, with greater vision, with a finer spirit of hope and achievement. You are here to enrich the world, and you impoverish yourself if you forget the errand."
- *M.D. Arnold:* "A good leader leads the people from above them. A great leader leads the people from within them."
- *Ronald Reagan:* "The greatest leader is not necessarily the one who does the greatest things. He is the one that gets the people to do the greatest things."

Of course, the problem is that many who occupy positions of authority within any hierarchy do not conform to such ideals but, being their own unique selves, pursue their private aspirations and predilections with such vigor they sometimes fully deserve the utter disdain Mark Twain supposedly spewed upon certain politicians: *"Politicians and diapers must be changed often, and for the same reason."*

I say 'supposedly' because many authorities consider this quote misattributed to Twain… that it is actually a recent fabrication of social media. I find that very interesting! Whatever, the quote itself is obviously quite popular in contemporary America.

Well, what politician, business executive, or any other public administrator has not run the gamut of eloquent praise and gutter-vituperation? It's an American tradition to keep-at-ready a dose of invective; a tradition that goes back to the eighteenth Century during the travail between England and its American colonies regarding the issue of 'taxation-without representation'. In a desperate effort to convince King George III to cease his disregard of their rights as English citizens, the colonies sent him a letter that not only described their extreme displeasure but also fairly hinted that they intended to take appropriate action if he ignored their complaint.

> **An Address to the People of Great Britain Friends and Fellow Subjects**
>
> When a Nation, led to greatness by the hand of liberty, and blessed by all the Glory that heroism, munificence, and humanity can bestow, descends to the ungrateful task of forging chains for her friends and children, and instead of giving support to Freedom, turns advocate for slavery and oppression, there is reason to suspect that she has either ceased to be virtuous or been extremely negligent in the appointment of her Rulers. (Ellis, 2021a, b, from National Archives)

King George ignored them. The colonies responded with a seven-year rebellion that ended with the creation of their new Nation, the United States. Maybe the king learned a lesson, but I doubt it. After all, leaders are as unique as any other human, subject (as we all are) to the same convolutional influences of personal prejudices, subconscious inclinations, and private aspirations. We humans not only argue with each other – we argue with ourselves!

While variants of social structure have evolved over many millennia, I doubt that the multifarious dynamics of the human brain have changed in any discernable way. Thus, leadership in any hierarchically structured, democratic society composed of millions is the bottom-line tool for fashioning consensus out of chaos. But, is leadership defined solely by a mastery of building consensus? What of character?

> If your actions inspire others to dream more, learn more, do more and become more, you are a leader. (John Quincy Adams)[4]

Our long-ago ancestors daily depended on the skills of temporary leaders. In the modern world, we depend on both the *skills* and the *character* of professional leaders.

In the U.S., there are many and diverse governmental and public-sector organizations (e.g. business schools, medical care facilities, public-relations management, etc.) and that provide not only training in conflict-resolution, but also certification that is increasingly a requirement for employment-advancement. Such training may involve classroom or on-the-job participation, or attendance at professional conferences; it is also available on-line.

11.2 Well, So What?

It might appear that I have been painting a rather gloomy picture of hierarchical decision-making. Quite the contrary. I have merely described some of its obvious dynamics as determined by the uniqueness of each individual within a large population having easy access to instantaneous and world-wide communication-technologies. Recent misuses of those technologies in interpersonal bullying, political disarray, and pandemic-response are obvious, but there are also positive outcomes that must be given serious attention – as the old saying goes *"When life gives you lemons, make lemonade"*.

For example, recent TV-adds have shown a couple of lovers on their way to a restaurant; the male is substantially shorter than the female, or they are different colors or of different ethnicities. In some adds, the couple is of the same gender and are shown kissing. Such scenes would never have been allowed 20 years ago. Things are changing; old stereotypes based on prejudice have no relevance to social reality; they are old, worn-out artifacts that have no place in the twenty-first Century.

Of course, there are pockets of resistance. I recently watched a TV-add for a law-firm. It included a view of the firm's thirty lawyers all lined up and smiling over a subtitle "40 years of experience". The lawyers were all white; there was one female. The subtitle should have read "40 years out of date". Little did the designer of that add realize that what he/she was really advertising was the likely demise of that law firm.

Yes, the uniqueness of each human, the increase in population density, and the ever-increasing availability of advanced communication-technology have interacted to foster individual harm and social discord, but in so doing, have also exposed previously neglected opportunities and the means for effecting needed and desirable change.

---

[4] Sixth President of U.S. (1825–1829), Diplomat, Senator, Secretary of State, member of Federal and State Houses of Representatives. He devoted his entire life to National service.

A constant factor in human history is CHANGE driven by the variants of threat, adversity, and calamity. Despite the carnage that have ensued from each, each has also resulted in new ways of thinking and behavior, as in (previously discussed) pandemic-related changes in productivity using virtual technology, and climate-related changes in agriculture and the ongoing replacement od gas-powered with electric-powered vehicles.

Typically, any menace that humans have faced has (hopefully sooner or (sometimes later) resulted in inventive, necessary, and effective change.

However, in a democratic population of millions, real changes are almost impossible without significant popular support. For example, while President Kennedy proposed a civil-rights agenda in 1963, the mass-media coverage of his assassination allowed President Johnson, *with the vast support of a stunned nation*, to effect congressional enactment of the Civil-Rights Act of 1964 despite previous viperous congressional opposition.

Of course, popular support of inaction may also constitute a bedrock-foundation of stagnation – that familiar, warm comfort of status quo.

That's a fundamental function of democratic leadership: not to dictate, but to help the public resolve the quandary of Shakespeare's Hamlet:

> To be, or not to be, that is the question:
> Whether 'tis nobler in the mind to suffer'
> The slings and arrows of outrageous fortune,
> Or to take arms against a sea of troubles
> And by opposing end them.

I trust you will overlook your likely chagrin of my introducing poetry in this discussion, but I merely want to make a particular point by translating Shakespeare's eloquence into contemporary English:

1. "*nobler in the mind*": best and admirable, desirable, ethical, and moral decision
2. "*suffer*": put up with, ignore, withstand
3. "*slings, arrows and outrageous fortune*": threats, dangers, catastrophes, emergencies
4. "*to take arms*": to do something about it, to fix it, make it better

The point? What Shakespeare wrote at the very beginning of the seventeenth Century is directly applicable to emergency planning and response in the twenty-first Century. We can either ignore our past failures or correct them by eliminating inequities born of prejudice and/or outdated tradition and habit and employing new technologies and new understandings of the dynamic interactions, indeed, the interdependence of Humanity and Nature.

The need for change in our thinking about and responding to emergencies is probably best made evident by the Category 5 Atlantic Hurricane Katrina (late August, 2005) which, with maximum winds of 174 mph, resulted in over 1800 fatalities and $126 billions in structural damage. Both during and after Katrina, response efforts were severely criticized by American as well as by foreign journalists, and Federal, State, and Municipal politicians. Major assessments of the overall response were separately compiled by a Presidentially appointed committee and a

Select Bipartisan Congressional Committee;[5] these assessments resulted in the identification of numerous, and substantive changes required in the execution of future response efforts. While many may consider these volumes, in combination with previous (if not on-going) complaints disseminated by mass-social media, evidentiary proof of incompetence, I strongly disagree. We learn through our mistakes.

Any failure of the moment spotlights potential pathways to future success. Every advance in science (for example) derives from previous error; incompetence in any task undertaken by humans is evident only by our having ignored previous error. This axiom is integral to our understanding that any plan is a temporary guide. Every emergency, even those of the same kind, shows its own distinctiveness, whether in magnitude, extent, duration, and its timely coincidence with unpredictable circumstances contrived by humans and/or Nature.

> *A plan is a continuous, evolving instrument of anticipated actions that maximize opportunities and guide response operations. Since planning is an ongoing process,* a plan is an interim product based on information and understanding at the moment and is subject to revision. *That is why plans are best described as 'living' documents.* (Department of Homeland Security, 2008)[6]

Given the utter complexity involved in implementing any plan of response, it should not come as a surprise that logistical problems receive primary attention not only of responders but also of the mass media and, subsequently, the public. Less obvious are the percipient deficiencies of the plan; it is easier to recognize failures related to the transport of needed materials and vehicles and the availability of temporary shelter than to recognize failures related to the holistic comprehensiveness – the scope – of the plan itself.

Today, the incipient threat of global climate change has become obvious; it is a reality that promises to be (excuse the vulgate) the 'grandaddy' of emergencies experienced by modern global-civilization and, thus, a thoroughly torturous test of the adage "all for one; one for all". In consequence, some nations are experimenting with new ways of thinking that suggest if not preventive, at least ameliorative behavior. In the U.S., the accelerated substitution of electrically-powered for petroleum-powered vehicles is one example. In Finland, a confederation of governmental and industrial authorities, motivated by climate change and diminishing natural resources has instituted what is called a "circular economy" which does away with the very concept of waste. Many other nations are expressing increased interest in this new approach.

Well, these are but two examples of new approaches that demonstrate the need for new thinking as the necessary prelude to new and highly effective new responses

[5] "The Federal Response to Hurricane Katrina: Lessons Learned", The White House, February 2006.

"A Failure of Initiative: Final Report of the Select Bipartisan Committee to Investigate the Preparation for and Response to Hurricane Katrina", 2006. (***note:*** print volumes available via the U.S. Printing Office, or may be downloaded from the Internet)

[6] Underline emphasis added by this author; this document (National Response Framework) and subsequent editions by the U.S. Department of Homeland Security are available via the Internet.

to threats. Change is necessary, even if always difficult. It is time we forgo our contentment with what has become robotic-routine and move our thinking about any emergency into the twenty-first Century.

Suggested Exercise for Chapter 11

Select any one type of emergency discussed in Part II of this volume and, having researched an historical example, identify, and discuss in an oral or written presentation what you consider to be significant failures regarding:

- Prevention of the emergency
- Comprehensiveness of response regarding societal distribution of consequences
- Duration of response
- Attention given to that emergency as becoming the source of subsequent emergencies in the same local and/or distant location
- Attention given to long term effects on first-responders

References

Crosby, A. (2003). *America's forgotten pandemic: The influenza of 1918*. Cambridge University Press.

Ellis, J. J. (2021a). *The cause: The American revolution and it's discontents, 1773–1783*. W. W. Norton & Company.

Ellis, J. J. (2021b). *The cause: The American revolution and it's discontents, 1773–1783*. W. W. Norton & Company. (Original source: An Address to the People of Great Britain, American Archives).

Gavrilets, S., et al. (2016). Convergence to consensus in heterogeneous groups and the emergence of informal leadership. *Scientific Reports, 6*, 29704. https://doi.org/10.1038/srep29704

Science News. (2022a, February 4). *Preventing pandemics costs far less than controlling them*. Duke University.

Science News. (2022b). *Climate change has likely begun to suffocate the word's fisheries*. American Geophysical Union.

Chapter 12
Epilogue

Previous chapters included discussions of a variety of issues. I think it appropriate to summarize these issues in a less discursive manner – call it a 'take-away message' that, though relatively brief, is a series of concise statements regarding what I have consistently referred to as a holistic approach to emergency planning and subsequent response.

1. The central sense of the word 'emergency' is that of *immediacy of responsive action*. However, the sense of 'immediacy' *does not define the duration* of that action. I have therefore suggested that we consider any emergency as an 'occurrence' (a 'happening' or 'situation') that requires immediate response. However, that individual "occurrence" typically results in a wide variety of both short- and long-term consequences. The covid pandemic, for example, has clearly resulted in a large variety of injury or other risks regarding death and health, but also long-term effects on industrial productivity, personal and institutional finance, education, politics, personal and social psychology, and even global interactions. Each of such (and other) effects are the real emergencies.
2. The consequences of a particular occurrence are distributed unevenly among *diverse Individuals or groups of humans and the environment.* Human victims include those who are exposed by chance and or by social inequities (e.g., color, gender, income, etc.) or by societal designation as first-responders or essential personnel (e.g., medical personnel and other municipal personnel). Environmental victims include biota and habitats exposed by chance but, also by human behavior (e.g., wildfires begun on purpose or by the carelessness of humans).
3. Any occurrence involves the complex interplay between Nature's dynamics and human behavior. We cannot change Nature's dynamics, but we can change human behavior. For example, why do we continue to spend billions of dollars responding to wildfires, the majority of which are caused by humans? Why do we keep expanding cities into areas known to be prone to terrestrial earthquakes or to tsunami devastation by oceanic quakes?

P. A. Erickson, *Effective Environmental Emergency Responses*,
https://doi.org/10.1007/978-3-031-05893-6_12

4. Of course, changing human behavior depends heavily upon building a public consensus. Achieving a consensus within a modern society requires (a) leadership and (b) appropriate and persistent utilization of mass and social media. Some political leaders cater primarily to the prejudices of their electorate rather than risk alienating them (and, thus, threatening their own job security) by guiding them toward more desirable societal objectives.

Regarding the purveyance of defamation, lies and misinformation by social media, much attention is currently given to the proper means of regulation. As for the mass media, it is important to realize that for two decades, Walter Cronkite[1] was known as "the most trusted man in America". Well, in those days, National news was dominated by professional journalists; today, we have an increasing infusion of broadcasters (maybe better described as 'Personalities') who have a bent for providing entertainment rather than an understanding of newsworthy information, a situation that is beginning to tickle the nerves of regulatory concern within professional journalism.

Historically, the American approach to emergencies has been a coordinated response of Federal and State agencies that necessarily focuses on logistics, the appropriate equipment, supplies, and personnel to achieve a "return to normalcy". But what does normalcy mean? A return to the daily routines that existed before the "occurrence" that forced a response? If so, just whose routines are we talking about?

Any executed plan for action has its logistical failures, be they failures due to chance, to errors in decision and/or poor communication between individual members of a chain of command. Such errors occur in the execution of any plan, whether that plan is devised for responding to an emergency, or placing a satellite in a specific orbit, or arranging a celebratory gathering. Errors in the scope (comprehensiveness) of any plan are quite different. Those are errors primarily due to human deficiencies in thoughtfulness.

Evacuate immediately! But what of those who do not have personal vehicles?
Send to hospital! But what of those who do not have health insurance?
Evacuate homes on Pine View Ave! But are there any homeless in adjacent areas?
Send evacuees to temporary shelter! But does that shelter have bathroom facilities, lights, food, and police protection?

HEADLINE NEWS: WILDFIRE FINALLY OUT! But could it have been prevented?
HEADLINE NEWS (3 months later): THINGS BACK TO NORMAL! Really? For everyone?

Hell no! After all, each day, we do many things that we normally do...... in the sense we usually do the same. But every tomorrow has its own state of normality. Every tomorrow is somehow different. Despite your most deep desire, there is no return to yesterday's normal.

Throughout this volume I have emphasized that emergency planning and subsequent response is an appropriate measure of purported National values and have suggested that the scope of emergency planning be expanded to minimize threats derived from the interactions of human behavior and environmental dynamics, to

[1] Journalist; anchorman for CBS Evening News 1962–1981.

eradicate those threats derived from persistent social-inequities, and to maximize our use of ever-expanding sciences to replace out-of-date traditions with more enlightened understandings. The latter is most difficult as is made obvious by the following quote from a news article I read only a few hours before writing this sentence.

> **Washington Post By *Joel Achenbach* February 17, 2022 at 1:00 p.m. EST**
>
> Many Americans … blame public health proponents for prolonging the crisis and preventing the return to normal. That leads to the next turn of the screw: If a lingering effect of the pandemic is that people view public health interventions as overreach, or somehow corrupted by factors other than saving lives, the country will be in worse shape the next time a virus comes roaring out of nowhere, sickening people en masse.
>
> That's the conclusion of Andrew Noymer, a University of California at Irvine epidemiologist, who, when asked what the big lesson of the pandemic is, replied in an email that "public health as an intellectual endeavor is weak; its own subject-matter experts get swept aside in a crisis," and "we are less, not more, prepared for the next pandemic, after this."
>
> Much of the rancor and unhappiness dates to early in the pandemic – the revolt against masking, led by red state Republican officials and libertarians generally – was well underway by summer 2020. But lately, some of the calls of no-more-restrictions are coming from inside the house: Many former supporters of pandemic interventions have moved to the other side, saying the lower caseloads and availability of vaccines and therapeutics mean it's time to fully reopen society – even if the CDC continues to *endorse indoor masking* in schools and in areas with substantial or high coronavirus transmission.
>
> "Open Everything" declared *an article* in the Atlantic by Yascha Mounk, a former supporter of interventions. An organization called Urgency of Normal is pushing for an end to mask mandates in schools, saying in its *mission statement* that continued pandemic restrictions are now a greater threat to students than the virus: "Children – and their parents – have shouldered an outsize burden long enough. Restoring normal childhood is a moral imperative, based on the balance of today's evidence."
>
> Some disease experts bristle at that argument, saying it reflects the views of healthy people in privileged strata of society. In more vulnerable areas, schools lack good ventilation, vaccine uptake is relatively low, and students are more likely to live in multigenerational families with elderly members who have a higher risk of severe disease from the virus.
>
> Gonsalves expressed that in pungent terms in a long Twitter thread recently:
>
> "This isn't my first time at the rodeo. With HIV, I saw how very privileged people were willing to f--- over others and let a virus flourish in the US and around the world, once they personally had access to potent antiretroviral drugs," he wrote.
>
> Gounder said she suspects some of the pushback against restrictions comes from vaccinated people who think they earned a return to normalcy: "I think people feel like, 'I followed the rules, I should be able to get on with my life.'"

Well, I hope I have succeeded in at least tickling those 'better angels' in you…… waking them up to this new century which, as young as it is, presents new opportunities and the means for transforming easy and empty expressions of sympathy into meaningful action. If not …… what the heck! I tried. Meanwhile, my best wishes for you and yours… and for everyone.

Appendices

Appendix 1

Appendix 1A: NIH National Institute of Mental Health *Transforming the Understanding and Treatment of Mental Illness*

Post-Traumatic Stress Disorder

Overview

Post-traumatic stress disorder (PTSD) is a disorder that develops in some people who have experienced a shocking, scary, or dangerous event.

It is natural to feel afraid during and after a traumatic situation. Fear triggers many split-second changes in the body to help defend against danger or to avoid it. This "fight-or-flight" response is a typical reaction meant to protect a person from harm. Nearly everyone will experience a range of reactions after trauma, yet most people recover from initial symptoms naturally. Those who continue to experience problems may be diagnosed with PTSD. People who have PTSD may feel stressed or frightened, even when they are not in danger.

Signs and Symptoms

While most but not all traumatized people experience short term symptoms, the majority do not develop ongoing (chronic) PTSD. Not everyone with PTSD has been through a dangerous event. Some experiences, like the sudden, unexpected death of a loved one, can also cause PTSD. Symptoms usually begin early, within 3 months of the traumatic incident, but sometimes they begin years afterward. Symptoms must last more than a month and be severe enough to interfere with

P. A. Erickson, *Effective Environmental Emergency Responses*,
https://doi.org/10.1007/978-3-031-05893-6

relationships or work to be considered PTSD. The course of the illness varies. Some people recover within 6 months, while others have symptoms that last much longer. In some people, the condition becomes chronic.

A doctor who has experience helping people with mental illnesses, such as a psychiatrist or psychologist, can diagnose PTSD.

To be diagnosed with PTSD, an adult must have all of the following for at least 1 month:

- At least one re-experiencing symptom
- At least one avoidance symptom
- At least two arousal and reactivity symptoms
- At least two cognition and mood symptoms

Re-experiencing symptoms include:

- Flashbacks—reliving the trauma over and over, including physical symptoms like a racing heart or sweating
- Bad dreams
- Frightening thoughts

Re-experiencing symptoms may cause problems in a person's everyday routine. The symptoms can start from the person's own thoughts and feelings. Words, objects, or situations that are reminders of the event can also trigger re-experiencing symptoms.

Avoidance symptoms include:

- Staying away from places, events, or objects that are reminders of the traumatic experience
- Avoiding thoughts or feelings related to the traumatic event

Things that remind a person of the traumatic event can trigger avoidance symptoms. These symptoms may cause a person to change his or her personal routine. For example, after a bad car accident, a person who usually drives may avoid driving or riding in a car.

Arousal and reactivity symptoms include:

- Being easily startled
- Feeling tense or "on edge"
- Having difficulty sleeping
- Having angry outbursts

Arousal symptoms are usually constant, instead of being triggered by things that remind one of the traumatic events. These symptoms can make the person feel stressed and angry. They may make it hard to do daily tasks, such as sleeping, eating, or concentrating.

Cognition and mood symptoms include:

- Trouble remembering key features of the traumatic event
- Negative thoughts about oneself or the world

- Distorted feelings like guilt or blame
- Loss of interest in enjoyable activities

Cognition and mood symptoms can begin or worsen after the traumatic event, but are not due to injury or substance use. These symptoms can make the person feel alienated or detached from friends or family members.

It is natural to have some of these symptoms for a few weeks after a dangerous event. When the symptoms last more than a month, seriously affect one's ability to function, and are not due to substance use, medical illness, or anything except the event itself, they might be PTSD. Some people with PTSD don't show any symptoms for weeks or months. PTSD is often accompanied by depression, substance abuse, or one or more of the other anxiety disorders.

Do Children React Differently Than Adults?

Children and teens can have extreme reactions to trauma, but some of their symptoms may not be the same as adults. Symptoms sometimes seen in very young children (less than 6 years old), these symptoms can include:

- Wetting the bed after having learned to use the toilet
- Forgetting how to or being unable to talk
- Acting out the scary event during playtime
- Being unusually clingy with a parent or other adult

Older children and teens are more likely to show symptoms similar to those seen in adults. They may also develop disruptive, disrespectful, or destructive behaviors. Older children and teens may feel guilty for not preventing injury or deaths. They may also have thoughts of revenge.

Risk Factors

Anyone can develop PTSD at any age. This includes war veterans, children, and people who have been through a physical or sexual assault, abuse, accident, disaster, or other serious events. According to the National Center for PTSD, about 7 or 8 out of every 100 people will experience PTSD at some point in their lives. Women are more likely to develop PTSD than men, and genes may make some people more likely to develop PTSD than others.

Not everyone with PTSD has been through a dangerous event. Some people develop PTSD after a friend or family member experiences danger or harm. The sudden, unexpected death of a loved one can also lead to PTSD.

Why Do Some People Develop PTSD and Other People Do Not?

It is important to remember that not everyone who lives through a dangerous event develops PTSD. In fact, most people will not develop the disorder.

Many factors play a part in whether a person will develop PTSD. Some examples are listed below. *Risk factors* make a person more likely to develop PTSD. Other factors, called *resilience factors*, can help reduce the risk of the disorder.

Some factors that increase risk for PTSD include:

- Living through dangerous events and traumas
- Getting hurt
- Seeing another person hurt, or seeing a dead body
- Childhood trauma
- Feeling horror, helplessness, or extreme fear
- Having little or no social support after the event
- Dealing with extra stress after the event, such as loss of a loved one, pain and injury, or loss of a job or home
- Having a history of mental illness or substance abuse

Some factors that may promote recovery after trauma include:

- Seeking out support from other people, such as friends and family
- Finding a support group after a traumatic event
- Learning to feel good about one's own actions in the face of danger
- Having a positive coping strategy, or a way of getting through the bad event and learning from it
- Being able to act and respond effectively despite feeling fear

Researchers are studying the importance of these and other risk and resilience factors, including genetics and neurobiology. With more research, someday it may be possible to predict who is likely to develop PTSD and to prevent it.

Treatments and Therapies

The main treatments for people with PTSD are medications, psychotherapy ("talk" therapy), or both. Everyone is different, and PTSD affects people differently, so a treatment that works for one person may not work for another. It is important for anyone with PTSD to be treated by a mental health provider who is experienced with PTSD. Some people with PTSD may need to try different treatments to find what works for their symptoms.

If someone with PTSD is going through an ongoing trauma, such as being in an abusive relationship, both of the problems need to be addressed. Other ongoing problems can include panic disorder, depression, substance abuse, and feeling suicidal.

Medications

The most studied type of medication for treating PTSD are antidepressants, which may help control PTSD symptoms such as sadness, worry, anger, and feeling numb inside. Other medications may be helpful for treating specific PTSD symptoms, such as sleep problems and nightmares.

Doctors and patients can work together to find the best medication or medication combination, as well as the right dose. Check the U.S. Food and Drug Administration website for the latest information on patient medication guides, warnings, or newly approved medications.

Psychotherapy

Psychotherapy (sometimes called "talk therapy") involves talking with a mental health professional to treat a mental illness. Psychotherapy can occur one-on-one or in a group. Talk therapy treatment for PTSD usually lasts 6–12 weeks, but it can last longer. Research shows that support from family and friends can be an important part of recovery.

Many types of psychotherapy can help people with PTSD. Some types target the symptoms of PTSD directly. Other therapies focus on social, family, or job-related problems. The doctor or therapist may combine different therapies depending on each person's needs.

Effective psychotherapies tend to emphasize a few key components, including education about symptoms, teaching skills to help identify the triggers of symptoms, and skills to manage the symptoms. One helpful form of therapy is called cognitive behavioral therapy, or CBT. CBT can include:

- **Exposure therapy.** This helps people face and control their fear. It gradually exposes them to the trauma they experienced in a safe way. It uses imagining, writing, or visiting the place where the event happened. The therapist uses these tools to help people with PTSD cope with their feelings.
- **Cognitive restructuring.** This helps people make sense of the bad memories. Sometimes people remember the event differently than how it happened. They may feel guilt or shame about something that is not their fault. The therapist helps people with PTSD look at what happened in a realistic way.

There are other types of treatment that can help as well. People with PTSD should talk about all treatment options with a therapist. Treatment should equip individuals with the skills to manage their symptoms and help them participate in activities that they enjoyed before developing PTSD.

How Talk Therapies Help People Overcome PTSD

Talk therapies teach people helpful ways to react to the frightening events that trigger their PTSD symptoms. Based on this general goal, different types of therapy may:

- Teach about trauma and its effects
- Use relaxation and anger-control skills
- Provide tips for better sleep, diet, and exercise habits
- Help people identify and deal with guilt, shame, and other feelings about the event
- Focus on changing how people react to their PTSD symptoms. For example, therapy helps people face reminders of the trauma.

Beyond Treatment: How Can I Help Myself?

It may be very hard to take that first step to help yourself. It is important to realize that although it may take some time, with treatment, you can get better. If you are unsure where to go for help, ask your family doctor. You can also check NIMH's Help for Mental Illnesses page or search online for "mental health providers," "social services," "hotlines," or "physicians" for phone numbers and addresses. An emergency room doctor can also provide temporary help and can tell you where and how to get further help.

To help yourself while in treatment:

- Talk with your doctor about treatment options
- Engage in mild physical activity or exercise to help reduce stress
- Set realistic goals for yourself
- Break up large tasks into small ones, set some priorities, and do what you can as you can
- Try to spend time with other people, and confide in a trusted friend or relative. Tell others about things that may trigger symptoms.
- Expect your symptoms to improve gradually, not immediately
- Identify and seek out comforting situations, places, and people

Caring for yourself and others is especially important when large numbers of people are exposed to traumatic events (such as natural disasters, accidents, and violent acts).

Next Steps for PTSD Research

In the last decade, progress in research on the mental and biological foundations of PTSD has lead scientists to focus on better understanding the underlying causes of why people experience a range of reactions to trauma.

- NIMH-funded researchers are exploring trauma patients in acute care settings to better understand the changes that occur in individuals whose symptoms improve naturally.
- Other research is looking at how fear memories are affected by learning, changes in the body, or even sleep.
- Research on preventing the development of PTSD soon after trauma exposure is also under way.
- Other research is attempting to identify what factors determine whether someone with PTSD will respond well to one type of intervention or another, aiming to develop more personalized, effective, and efficient treatments.
- As gene research and brain imaging technologies continue to improve, scientists are more likely to be able to pinpoint when and where in the brain PTSD begins. This understanding may then lead to better targeted treatments to suit each person's own needs or even prevent the disorder before it causes harm.

Last Revised: May 2019

Appendix 1B: Transforming the Understanding and Treatment of Mental Illnesses

Generalized Anxiety Disorder: When Worry Gets Out of Control

What Is GAD?

Occasional anxiety is a normal part of life. You might worry about things like health, money, or family problems. But people with generalized anxiety disorder (GAD) feel extremely worried or feel nervous about these and other things—even when there is little or no reason to worry about them. People with GAD find it difficult to control their anxiety and stay focused on daily tasks.

The good news is that GAD is treatable. Call your doctor to talk about your symptoms so that you can feel better.

What Are the Signs and Symptoms of GAD?

GAD develops slowly. It often starts during the teen years or young adulthood. People with GAD may:

- Worry very much about everyday things
- Have trouble controlling their worries or feelings of nervousness
- Know that they worry much more than they should
- Feel restless and have trouble relaxing
- Have a hard time concentrating
- Be easily startled
- Have trouble falling asleep or staying asleep
- Feel easily tired or tired all the time
- Have headaches, muscle aches, stomach aches, or unexplained pains
- Have a hard time swallowing
- Tremble or twitch
- Be irritable or feel "on edge"
- Sweat a lot, feel light-headed or out of breath
- Have to go to the bathroom a lot

Children and teens with GAD often worry excessively about:

- Their performance, such as in school or in sports
- Catastrophes, such as earthquakes or war

Adults with GAD are often highly nervous about everyday circumstances, such as:

- Job security or performance
- Health
- Finances
- The health and well-being of their children
- Being late
- Completing household chores and other responsibilities

Both children and adults with GAD may experience physical symptoms that make it hard to function and that interfere with daily life.

Symptoms may get better or worse at different times, and they are often worse during times of stress, such as with a physical illness, during exams at school, or during a family or relationship conflict.

What Causes GAD?

GAD sometimes runs in families, but no one knows for sure why some family members have it while others don't. Researchers have found that several parts of the brain, as well as biological processes, play a key role in fear and anxiety. By learning more about how the brain and body function in people with anxiety disorders, researchers may be able to create better treatments. Researchers are also looking for ways in which stress and environmental factors play a role.

How Is GAD Treated?

First, talk to your doctor about your symptoms. Your doctor should do an exam and ask you about your health history to make sure that an unrelated physical problem is not causing your symptoms. Your doctor may refer to you a mental health specialist, such as a psychiatrist or psychologist.

GAD is generally treated with psychotherapy, medication, or both. Talk with your doctor about the best treatment for you.

Psychotherapy

A type of psychotherapy called cognitive behavioral therapy (CBT) is especially useful for treating GAD. CBT teaches a person different ways of thinking, behaving, and reacting to situations that help him or her feel less anxious and worried. For more information on psychotherapy, visit http://www.nimh.nih.gov/health/topics/psychotherapies

Medication

Doctors may also prescribe medication to help treat GAD. Your doctor will work with you to find the best medication and dose for you. Different types of medication can be effective in GAD:

- Selective serotonin reuptake inhibitors (SSRIs)
- Serotonin-norepinephrine reuptake inhibitors (SNRIs)
- Other serotonergic medication
- Benzodiazepines

Doctors commonly use SSRIs and SNRIs to treat depression, but they are also helpful for the symptoms of GAD. They may take several weeks to start working. These medications may also cause side effects, such as headaches, nausea, or difficulty sleeping. These side effects are usually not severe for most people, especially if the dose starts off low and is increased slowly over time. **Talk to your doctor about any side effects that you have.**

Buspirone is another serotonergic medication that can be helpful in GAD. Buspirone needs to be taken continuously for several weeks for it to be fully effective.

Benzodiazepines, which are sedative medications, can also be used to manage severe forms of GAD. These medications are powerfully effective in rapidly decreasing anxiety, but they can cause tolerance and dependence if you use them continuously. Therefore, your doctor will only prescribe them for brief periods of time if you need them.

Don't give up on treatment too quickly. Both psychotherapy and medication can take some time to work. A healthy lifestyle can also help combat anxiety. Make sure to get enough sleep and exercise, eat a healthy diet, and turn to family and friends who you trust for support.

For basic information about these and other mental health medications, visit http://www. nimh.nih.gov/health/topics/mental-health-medications. Visit the Food and Drug Administration's website (http://www.fda.gov/) for the latest information on warnings, patient medication guides, or newly approved medications.

What Is It Like to Have GAD?

> I was worried all the time and felt nervous. My family told me that there were no signs of problems, but I still felt upset. I dreaded going to work because I couldn't keep my mind focused. I was having trouble falling asleep at night and was irritated at my family all the time.
>
> I saw my doctor and explained my constant worries. My doctor sent me to someone who knows about GAD. Now I am working with a counselor to cope better with my anxiety. I had to work hard, but I feel better. I'm glad I made that first call to my doctor.

Where Can I Find More Information?

Research, visit the NIMH website (http://www.nimh.nih.gov).

National Institute of Mental Health (NIMH)

Office of Science Policy, Planning, and Communications
Science Writing, Press, and Dissemination To learn more about generalized anxiety disorder, visit:

MedlinePlus (National Library of Medicine)

http://medlineplus.gov

(En español: http://medlineplus.gov/spanish)
For information on clinical trials, visit:

ClinicalTrials.gov

http://www.clinicaltrials.gov

(En español: http://salud.nih.gov/investigacion-clinica/)

For more information on conditions that affect mental health, resources, and Branch

6001 Executive Boulevard

Room 6200, MSC 9663

Bethesda, MD 20892-9663

Phone: 301-443-4513 or

1-866-615-NIMH (6464) toll-free

TTY: 301-443-8431 or

1-866-415-8051 toll free

Fax: 301-443-4279

Email: nimhinfo@nih.gov

Website: http://www.nimh.nih.gov

U.S. Department of Health and Human Services

National Institutes of Health

National Institute of Mental Health

NIH Publication No. 19-MH-8090

Revised 2016

Appendix 1C: Transforming the Understanding and Treatment of Mental Illnesses

Major Depression

Definitions

Major depression is one of the most common mental disorders in the United States. For some individuals, major depression can result in severe impairments that interfere with or limit one's ability to carry out major life activities.

Additional information can be found on the NIMH Health Topics page on Depression.

The past year prevalence data presented here for **major depressive episode** are from the 2017 National Survey on Drug Use and Health (NSDUH). The NSDUH study definition of major depressive episode is based mainly on the 5th edition of the Diagnostic and Statistical Manual of Mental Disorders (DSM-5):

- A period of at least 2 weeks when a person experienced a depressed mood or loss of interest or pleasure in daily activities, and had a majority of specified symptoms, such as problems with sleep, eating, energy, concentration, or self-worth.
- No exclusions were made for a major depressive episode symptoms caused by medical illness, substance use disorders, or medication.

Prevalence of Major Depressive Episode Among Adults

- Figure 1 shows the past year prevalence of major depressive episode among U.S. adults aged 18 or older in 2017.
  - An estimated 17.3 million adults in the United States had at least one major depressive episode. This number represented 7.1% of all U.S. adults.
  - The prevalence of major depressive episode was higher among adult females (8.7%) compared to males (5.3%).
  - The prevalence of adults with a major depressive episode was highest among individuals aged 18–25 (13.1%).
- The prevalence of major depressive episode was highest among adults reporting two or more races (11.3%).

Major Depressive Episode with Impairment Among Adults

- In 2017, an estimated 11 million U.S. adults aged 18 or older had at least one major depressive episode with severe impairment. This number represented 4.5% of all U.S. adults.
- Figure 2 shows overall past year prevalence of major depressive episode with and without severe impairment. Of adults with major depressive episode, 63.8% had severe impairment.

Treatment of Major Depressive Episode Among Adults

- Figure 3 shows data on treatment received within the past year by U.S. adults aged 18 or older with major depressive episode. Treatment types include health professional only, medication only, and health professional and medication combined.

*All other groups are non-Hispanic or Latino | **NH/OPI = Native Hawaiian / Other Pacific Islander| ***AI/AN = American Indian / Alaskan NativePercentPast Year Prevalence of Major Depressive Episode Among U.S. Adults (2017)Data Courtesy of SAMHSA7.17.18.78.75.3 5.313.113.17.77.74.74.75.45.47.97.95.45.44.44.44.74.78.08.011.3 11.3OverallFemaleSexMale18-25Age26-4950+Hispani…Race/EthnicityWhiteBlackAsianNH/OPI**AI/AN***2 or More02468101214

Past Year Prevalence of Any Mental Illness Among U.S. Adults (2017)		
Demographic		**Percent**
Overall		7.1
Sex	**Female**	8.7
	Male	5.3
Age	**18-25**	13.1
	26-49	7.7
	50+	4.7
Race/Ethnicity	**Hispanic or Latino***	5.4
	White	7.9
	Black	5.4
	Asian	4.4
	NH/OPI**	4.7
	AI/AN***	8.0
	2 or More	11.3

*All other groups are non-Hispanic or Latino | **NH/OPI = Native Hawaiian / Other Pacific Islander | ***AI/AN = American Indian / Alaskan Native

Fig. 1

- An estimated 65% received combined care by a health professional and medication treatment.
- Treatment with medication alone was least common (6%).
- Approximately 35% of adults with major depressive episode did not receive treatment.

Prevalence of Major Depressive Episode Among Adolescents

- Figure 4 shows the past year prevalence of major depressive episode among U.S. adolescents in 2017.
  - An estimated 3.2 million adolescents aged 12–17 in the United States had at least one major depressive episode. This number represented 13.3% of the U.S. population aged 12–17.

Past Year Severity of Major Depressive Episode Among U.S. Adults (2017)Data Courtesy of SAMHSAWithout SevereImpairment36%Without SevereImpairment36%With Severe Impairment64%With SevereImpairment64%

Past Year Severity of Major Depressive Episode Among U.S. Adults (2017)

Impairment Status	Percent
Without Severe Impairment	36.2
With Severe Impairment	63.8
Overall Prevalence	100

Fig. 2

Past Year Treatment Received Among Adults withMajor Depressive Episode (2017)Data Courtesy of SAMHSANo Treatment35%No Treatment35%Health Professional Only15%Health Professional Only15%Health ProfessionalAND Medication44%Health ProfessionalAND Medication44%Medication Only6%Medication Only6%

Past Year Treatment Received Among Adults with Major Depressive Episode (2017)

Treatment	Percent
No Treatment	35
Health Professional Only	15
Health Professional AND Medication	44
Medication Only	6
Total	100

Fig. 3

*All other groups are non-Hispanic or Latino / **AI/AN = American Indian/Alaska NativePercentPast Year Prevalence of Major Depressive Episode Among U.S. Adolescents (2017)Data Courtesy of SAMHSA13.313.320.020.06.86.84.84.88.88.811.811.817.217.216.9 16.918.518.513.813.814.014.09.59.511.311.316.316.316.9 16.9OverallFemaleSexMale12Age1314151617Hispa…Race/EthnicityWhiteBlackAsianAI/AN* *2 or …0510152025

Past Year Prevalence of Any Mental Illness Among U.S. Adults (2017)

Demographic		Percent
Overall		13.3
Sex	**Female**	20.0
	Male	6.8
Age	**12**	4.8
	13	8.8
	14	11.8
	15	17.2
	16	16.9
	17	18.5
Race/Ethnicity	**Hispanic***	13.8
	White	14.0
	Black	9.5
	Asian	11.3
	AI/AN**	16.3
	2 or more Races	16.9

*All other groups are non-Hispanic or Latino / **AI/AN = American Indian/Alaska Native

Fig. 4

- The prevalence of major depressive episode was higher among adolescent females (20.0%) compared to males (6.8%).
- The prevalence of major depressive episode was highest among adolescents reporting two or more races (16.9%).

Major Depressive Episode with Impairment Among Adolescents

- In 2017, an estimated 2.3 million adolescents aged 12–17 in the United States had at least one major depressive episode with severe impairment. This number represented 9.4% of the U.S. population aged 12–17.

- Figure 5 shows overall past year prevalence of major depressive episode with and without severe impairment among U.S. adolescents. Of adolescents with major depressive episode, approximately 70.77% had severe impairment.

Treatment of Major Depressive Episode Among Adolescents

- Figure 6 shows data on treatment received within the past year by U.S. adolescents aged 12–17 with major depressive episode in 2017. Treatment types included health professional only, medication only, and combined health professional and medication.
  - An estimated 19.6% received care by a health professional alone, and another 17.9% received combined care by a health professional and medication treatment.
  - Treatment with medication alone was least common (2.4%).
  - Approximately 60.1% of adolescents with major depressive episode did not receive treatment.

Population:

- The entirety of NSDUH respondents for the major depressive episode estimates is the civilian, non-institutionalized population aged 12–17 (adolescents) and 18 years old or older (adults) residing within the United States.

Past Year Severity of Major Depressive Episode Among U.S. Adolescents (2017)Data Courtesy of SAMHSAWithout severe impairment29%Without severe impairment29%With severe impairment71%With severe impairment71%

Past Year Severity of Major Depressive Episode Among U.S. Adolescents (2017)	
Impairment Status	**Percent**
Without severe impairment	29.3
With severe impairment	70.7
Overall Prevalence	100

Fig. 5

Past Year Treatment Received Among Adolescents withMajor Depressive Episode (2017)Data Courtesy of SAMHSANo Treatment60.1%No Treatment60.1%Health Professional Only19.6% Health Professional Only19.6%Health ProfessionalAND Medication17.9%Health Professional AND Medication17.9%Medication Only2.4%Medication Only2.4%

Past Year Treatment Received Among Adolescents with Major Depressive Episode (2017)	
Treatment	**Percent**
No Treatment	60.1
Health Professional Only	19.6
Health Professional AND Medication	17.9
Medication Only	2.4
Total	100

Fig. 6

- The survey covers residents of households (persons living in houses/townhouses, apartments, condominiums; civilians living in housing on military bases, etc.) and persons in non-institutional group quarters (e.g., shelters, rooming/boarding houses, college dormitories, migratory workers' camps, and halfway houses).
- The survey does not cover persons who, for the entire year, had no fixed address (e.g., homeless and/or transient persons not in shelters); were on active military duty; or who resided in institutional group quarters (e.g., correctional facilities, nursing homes, mental institutions, long-term hospitals).
- Some adults and adolescents in these excluded categories may have had a major depressive episode in the past year, but they are not accounted for in the NSDUH major depressive episode estimates.

Survey Non-response:

- In 2017, 32.9% of the selected NSDUH sample did not complete the interview.
- Reasons for non-response to interviewing include: refusal to participate (23.1%); respondent unavailable or no one at home (5.0%); and other reasons such as physical/mental incompetence or language barriers (4.8%).

- Adults and adolescents with major depressive episode may disproportionately fall into these non-response categories. While NSDUH weighting includes non-response adjustments to reduce bias, these adjustments may not fully account for differential non-response by mental illness status.

Please see the 2017 National Survey on Drug Use and Health Methodological Summary and Definitions report for further information on how these data were collected and calculated.

Last Updated: February 2019

Appendix 2

Appendix 2A

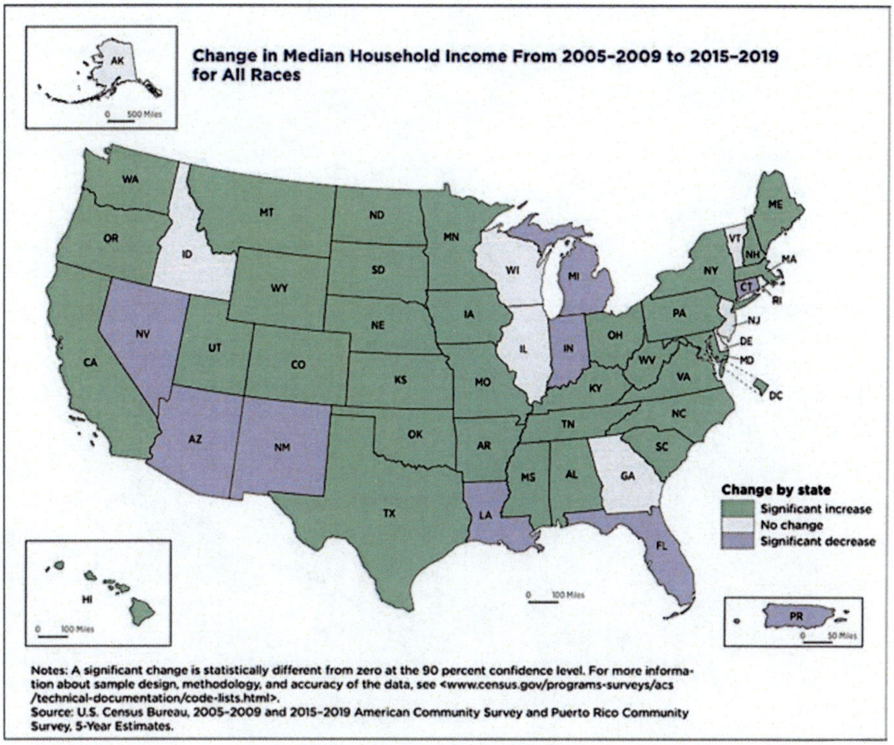

Appendix 2B

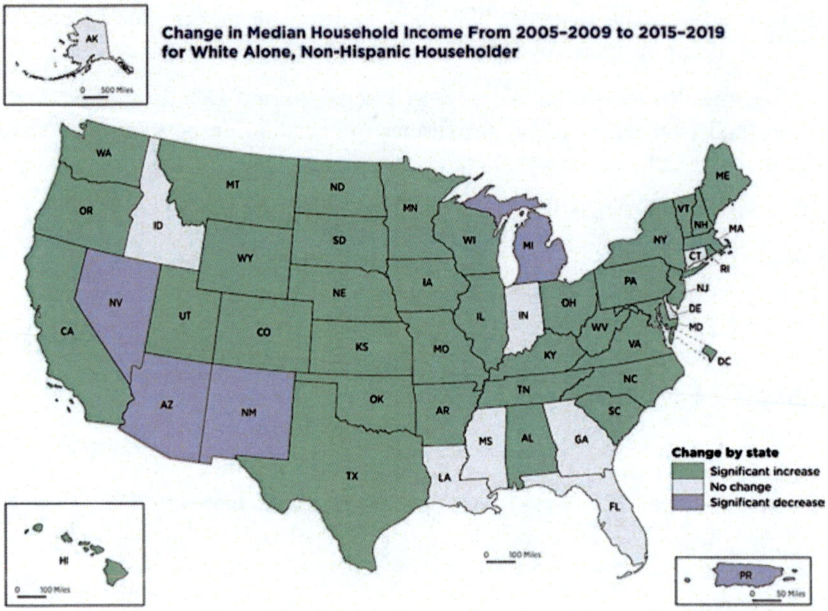

Appendix 2C

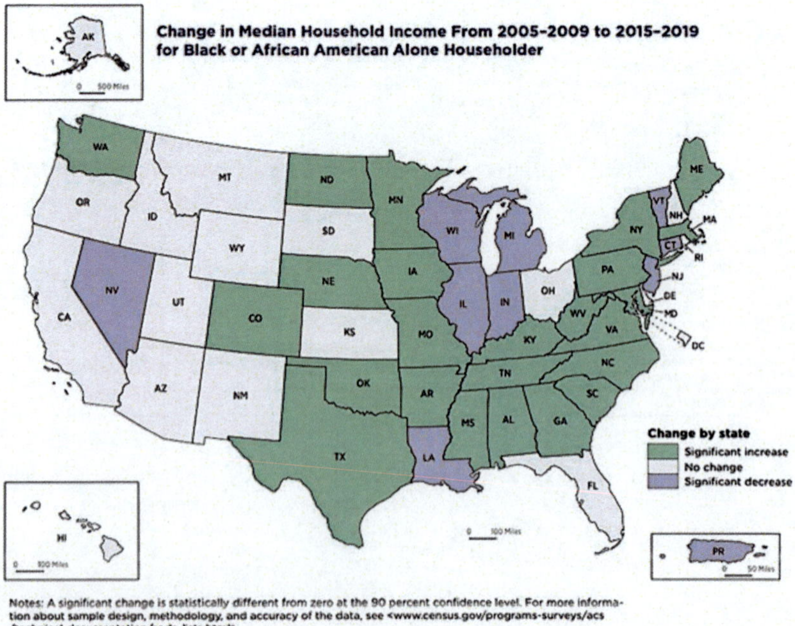

Notes: A significant change is statistically different from zero at the 90 percent confidence level. For more information about sample design, methodology, and accuracy of the data, see <www.census.gov/programs-surveys/acs/technical-documentation/code-lists.html>.
Source: U.S. Census Bureau, 2005–2009 and 2015–2019 American Community Survey and Puerto Rico Community Survey, 5-Year Estimates.

Appendix 2D

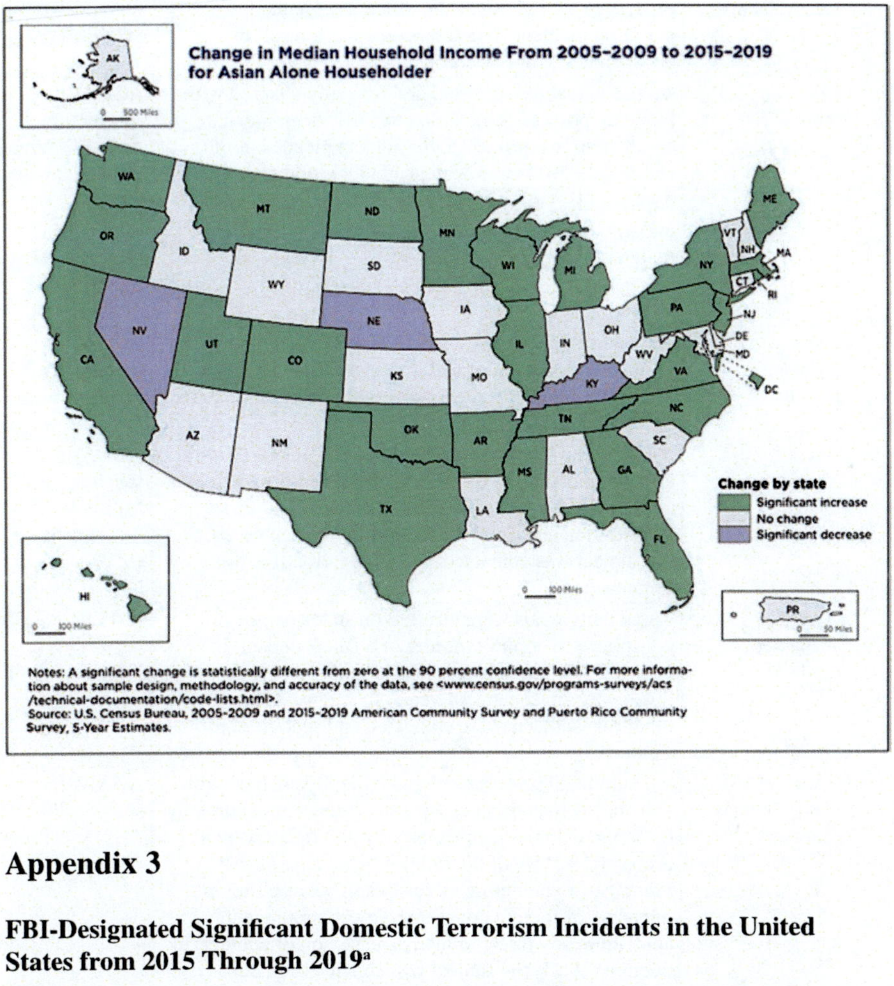

Appendix 3

FBI-Designated Significant Domestic Terrorism Incidents in the United States from 2015 Through 2019[a]

Date and location	Description	Category[b]
22 January 2015 Atlanta, Georgia	An individual was arrested and federally charged with tax evasion. The cumulative estimated losses totaled $1.5 million. In March 2016, the subject was sentenced to 1 year and 1 day.	Anti-Government or Anti-Authority Violent Extremism (AGAAVE), specifically Sovereign Citizen Violent Extremism (SCVE)
6 February 2015 Chicago, Illinois	An unidentified individual(s) allegedly set fire to and vandalized an occupied building containing horse carriages. The economic damages were estimated at $130,000.	Animal Rights/ Environmental Violent Extremism

(continued)

Date and location	Description	Category[b]
9 February 2015 Orange County, Florida	An individual shot and injured two law enforcement officers in an ambush as the officers responded to a domestic incident. The subject died as a result of engagement with law enforcement.	AGAAVE, specifically SCVE
14 February 2015 Elkins, West Virginia	An individual was arrested and federally charged with knowing possession of stolen explosives in interstate commerce. In April 2015, the subject pleaded guilty, and in July 2015, was sentenced to 8 months.	AGAAVE, specifically Anarchist Violent Extremism (AVE)
17 March 2015 Stover, Missouri	An individual was arrested and federally charged for making threats against the President of the United States.	Racially or Ethnically Motivated Violent Extremism (RMVE)
25 March 2015 St. Louis, Missouri	An individual was arrested and federally charged with knowing possession of a machinegun. In November 2015, the subject pleaded guilty, and in February 2016, was sentenced to 1 year and 1 day.	AGAAVE, specifically AVE
25 March 2015 Livingston, Montana	An individual was arrested and federally charged with knowing possession of a machinegun and possession of a firearm that is not registered in the National Firearms Registration and Transfer Record. In November 2015, the subject was convicted, and in March 2016, was sentenced to 6 years.	AGAAVE, specifically Militia Violent Extremism (MVE)
2 April 2015 Tallahassee, Florida	Three individuals were arrested on state charges of conspiracy to commit murder of a Black person. In March 2017, one subject pleaded guilty, and in April 2017, was sentenced to 4 years. In August 2017, the other two subjects were convicted and sentenced to 12 years.	RMVE
10 April 2015 Signal Mountain, Tennessee	An individual was arrested and federally charged with solicitation to commit a crime of violence; intentionally defacing, damaging, or destroying any religious real property, because of the religious character of that property, or attempting to do so; and transmitting in interstate or foreign commerce any communication containing any threat to injure the person of another. In February 2017, the subject was convicted. In September 2020, after going through a lengthy appellate process, the defendant was resentenced to 10 years.	AGAAVE, specifically MVE
17 June 2015 Charleston, South Carolina	An individual shot and killed nine Black people at the Emanuel AME Church. The subject was arrested on state charges of murder and was federally charged with hate crime acts resulting in death; hate crime act involving an attempt to kill; obstruction of free exercise of religious beliefs resulting in death; obstruction of free exercise of religious beliefs involving an attempt to kill and use of a dangerous weapon; and use of a firearm to commit murder during and in relation to a crime of violence. In December 2016, the subject was convicted, and in January 2017, was sentenced to death.	RMVE

(continued)

Date and location	Description	Category[b]
9 July 2015 Yavapai, Arizona	Two individuals were arrested on state charges for allegedly shooting and injuring a law enforcement officer during a routine traffic violation inquiry.	AGAAVE, specifically SCVE
31 July 2015 Vestavia Hills, Alabama	An individual was arrested on state charges for assaulting a Black convenience store clerk, and in May 2016, the subject pleaded guilty and received a 12 month suspended sentence.	RMVE
1 August 2015 Clover, South Carolina	Three individuals were arrested and federally charged with conspiracy to violate laws governing firearms and explosive devices. In September 2015, two subjects pleaded guilty, and in January 2016, were each sentenced to 21 months. In December 2015, the third subject pleaded guilty, and in June 2016, was sentenced to 22 months.	AGAAVE, specifically MVE
16 September 2015 Lawrence County, South Dakota	An individual stole a vehicle and shot at law enforcement officers during their pursuit of the subject. The subject died as a result of engagement with law enforcement.	Animal Rights/ Environmental Violent Extremism
23 September 2015 Wood County, West Virginia	An individual was arrested and convicted on state charges of threatening to commit a terrorist act after advocating to overthrow the government of West Virginia.	AGAAVE, specifically SCVE
7 October 2015 Richmond, Virginia	Three individuals were arrested and federally charged with conspiracy to commit robbery affecting commerce, felon in possession, and conspiracy to commit robbery affecting commerce for plotting to attack synagogues and Black churches. Each subject pleaded guilty and were later sentenced to 17.5 years, 8.75 years, and 7 years, respectively.	RMVE
27 November 2015 Colorado Springs, Colorado	An individual was arrested on state charges of first-degree murder for allegedly attacking a reproductive health care facility, shooting and killing three people and injuring others. The subject was later federally charged with violating the Freedom of Access to Clinic Entrances Act and use of a firearm during a crime of violence resulting in death where the killing is a murder. In total, three people were killed, including one law enforcement officer, and eight people were injured, including four law enforcement officers. The case is currently pending.	Abortion-Related Violent Extremism
2 January 2016 Harney County, Oregon	Multiple individuals seized and occupied the Malheur National Wildlife Refuge. In total, 18 subjects were sentenced to crimes associated with the occupation.	AGAAVE, specifically MVE and SCVE
19 February 2016 Nashville, Indiana	An individual was arrested by local authorities for allegedly attacking a Chinese student with a hatchet. In April 2016, a judge ruled the subject was not competent to stand trial.	RMVE
4 March 2016 Los Angeles, California	Three individuals were arrested on state charges of assault with a deadly weapon with special allegations of a hate crime for allegedly assaulting five Hispanic people.	RMVE

(continued)

Date and location	Description	Category[b]
1 May 2016 Seattle, Washington	An individual was arrested and federally charged with unlawful possession of destructive devices for throwing an improvised incendiary device at law enforcement officers, injuring one officer, at a May Day protest. In February 2018, the subject pleaded guilty, and in June 2018, was sentenced to 37 months.	AGAAVE, specifically AVE
18 June 2016 Salinas, Puerto Rico	An individual was arrested and federally charged with maliciously damaging or destroying, and attempting to damage or destroy, property by means of fire and explosives of a building and vehicle used in interstate commerce, for allegedly using incendiary devices to damage an agricultural business and targeting first responders, including law enforcements officers. In November 2016, the subject pleaded guilty, and in March 2017, was sentenced to 3 years.	Animal Rights/ Environmental Violent Extremism
22 June 2016 Mount Trumbull, Arizona	An individual was arrested and federally charged with attempted destruction of federal property by use of an explosive for plotting to destroy a Bureau of Land Management cabin. In April 2018, the subject pleaded guilty, and in July 2018, was sentenced to time served.	AGAAVE, specifically MVE
7 July 2016 Dallas, Texas	An individual shot at law enforcement officers during a protest. The subject shot and killed five law enforcement officers, injured six other officers, and injured two other people. The subject died as a result of engagement with law enforcement.	RMVE
7 July 2016 Bristol, Tennessee	An individual was arrested on state charges for shooting at passing cars on a highway and then at law enforcement officers who responded. The subject killed one person and injured one law enforcement officer. In December 2016, the subject pleaded guilty to state charges of first-degree murder and attempted first-degree murder and was sentenced to life.	RMVE
10 July 2016 Evansville, Indiana	An individual was arrested on state charges for the attempted murder of a transgender person. In February 2018, the subject pleaded guilty, and in May 2018, was sentenced to 36 years.	RMVE
17 July 2016 Baton Rouge, Louisiana	An individual shot at law enforcement officers, killing three officers and wounding six others. The subject died as a result of law enforcement engagement.	RMVE
1 August 2016 Randallstown, Maryland	An individual barricaded herself and her child in her residence when law enforcement officers attempted to serve a bench warrant for her failure to appear in court. The subject died as a result of law enforcement engagement.	AGAAVE, specifically SVCE
1 August 2016 Mahaska and Jasper Counties, Iowa	Individuals allegedly set fire to seven large vehicles used in the construction of the Dakota Access pipeline. Economic losses were estimated at approximately $3 million.	Animal Rights/ Environmental Violent Extremism

(continued)

Date and location	Description	Category[b]
16 August 2016 Olympia, Washington	An individual was arrested on state charges of assault in the second degree and malicious harassment for stabbing a Black member of an interracial couple. In October 2017, the subject pleaded guilty and was sentenced to 50 months.	RMVE
21 August 2016 Fort Wayne, Indiana	An individual was arrested on state murder charges for murdering a Black person. In July 2017, the subject pleaded guilty but mentally ill and was sentenced to 65 years.	RMVE
13 September 2016 Phoenix, Arizona	An individual was arrested on state charges of attempt to commit murder in the first degree, aggravated assault with a deadly weapon, endangerment, resisting arrest, and criminal damage for driving a vehicle into three white law enforcement officers, injuring two officers. In April 2019, the subject pleaded guilty, and in June 2019, was sentenced to 35 years.	RMVE
28 September 2016 Zionsville, Indiana	An individual was arrested on state charges of murder for shooting at law enforcement officers when they attempted to serve the arrest warrant for the murder of a white person. In February 2019, the subject was found guilty, and in April 2019, was sentenced to 65 years.	RMVE
30 September 2016 Van Buren Township, Michigan	An individual fled a routine traffic stop and led law enforcement officers on a high speed chase. The subject died as result of law enforcement engagement.	AGAAVE, specifically SCVE
7 October 2016 Hilliard, Ohio	An individual was arrested on state charges of conspiracy to commit murder for plotting to conduct a shooting at a high school. In January 2018, the subject plead no contest, and in February 2018, was sentenced to 4 years.	RMVE
4 October 2016 Garden City, Kansas	Three individuals were arrested for plotting a mass casualty attack using explosives against an apartment building where multiple Somali immigrants lived and worshipped. In January 2019, two subjects were found guilty of federal charges of conspiracy to use a weapon of mass destruction and conspiracy against civil rights, and the third subject was found guilty of knowingly and willingly making false statements. The subjects were sentenced to 30 years, 26 years, and 25 years, respectively.	AGAAVE, specifically MVE
16 October 2016 Jasper County, Iowa	Individuals allegedly set fire to four pieces of equipment used in the construction of the Dakota Access pipeline. Economic losses were estimated at approximately $2 million.	Animal Rights/ Environmental Violent Extremism
22 October 2016 La Pine, Oregon	An individual was arrested on state charges of reckless driving, third-degree escape, fourth-degree assault, resisting arrest, unlawful use of a weapon and reckless endangerment for assaulting a law enforcement officer, attempting to take the officer's weapon, and escaping custody during a vehicle pursuit. In October 2018, the subject pleaded guilty and was sentenced to 60 days.	AGAAVE, specifically SCVE

(continued)

Date and location	Description	Category[b]
27 October 2016 Morton County, North Dakota	Individuals allegedly participated in civil disorder, trespassing, and setting property on fire during a Dakota Access Pipeline protest. One subject was arrested and federally charged with civil disorder and use of fire to commit a federal felony crime. The subject later plead guilty to civil disorder, and in May 2018, was sentenced to 3 years.	Animal Rights/ Environmental Violent Extremism
8 November 2016 Buena Vista County, Iowa	Individuals allegedly set fire to four pieces of equipment used in the construction of the Dakota Access pipeline. Economic losses were estimated at approximately $2.5 million.	Animal Rights/ Environmental Violent Extremism
3 December 2016 Yanceyville, North Carolina	Two individuals were arrested on state charges of assault with a deadly weapon for allegedly stabbing and injuring their associate.	RMVE
1 February 2017 Berkeley, California	Individuals allegedly damaged property and assaulted law enforcement officers and attendees at an event featuring a political commentator.	AGAAVE, specifically AVE
15 February 2017 Myrtle Beach, South Carolina	An individual was arrested and federally charged with illegal possession of a firearm by a felon. In February 2018, the subject pleaded guilty, and in July 2018, was sentenced to 33 months.	RMVE
15 February 2017 Tipton County, Tennessee	Three individuals were arrested for conspiring to break an associate out of jail and kidnap a law enforcement officer and a judge. One subject was arrested by Canadian authorities in April 2017. The second subject pleaded guilty to federal charges of conspiracy to commit an act of violence, to wit, kidnapping, and in August 2017, was sentenced to 5 years. In December 2017, the third subject pleaded guilty to a federal charge of misprision of a felony and entered into a pre- trial diversion agreement.	AGAAVE, specifically SCVE
20 March 2017 New York City, New York	An individual was arrested on state charges of first-degree murder in furtherance of an act of terrorism, second-degree murder as a crime of terrorism, second-degree murder as a hate crime, and fourth-degree criminal possession of a weapon, for killing a Black person with a sword. In January 2019, the subject pleaded guilty, and in February 2019, was sentenced to life in prison.	RMVE
	An individual allegedly shot at law enforcement officers during the arrest of his father for alleged sovereign citizen fraudulent filings. The subject was injured as a result of law enforcement engagement.	AGAAVE, specifically SCVE
	An individual believed to be plotting a mass shooting was arrested on federal charges of illegal possession of a firearm. In September 2017, the subject pleaded guilty and was sentenced to 5 years of probation.	AGAAVE, specifically SCVE
	An individual was arrested on state charges of homicide, attempted homicide, discharging a firearm into an inhabited dwelling, and possession of a firearm by a felon for shooting and killing three white people.	RMVE

(continued)

Date and location	Description	Category[b]
	An individual allegedly purchased a fully automatic AK-47 and made statements about carrying out violence against law enforcement officers. The subject was arrested and federally charged with firearms offenses. In October 2018, the subject pleaded guilty; the case is pending.	RMVE
	Two individuals allegedly shot and killed a law enforcement officer during a traffic stop and then led officers on a high speed chase and shootout. One subject was arrested on state charges of deliberate homicide, and the other died as a result of engagement with law enforcement.	AGAAVE, specifically MVE
	An individual with a personalized violent ideology targeted and shot Republican members of Congress at a baseball field and wounded five people. The subject died as a result of engagement with law enforcement.	Domestic Violent Extremist (DVE)
	Three individuals were arrested and federally charged with unlawful possession of a machine gun, conspiracy to interfere with commerce by threats and violence, and attempted arson for allegedly bombing a mosque and a women's health clinic. In January 2019, two of the three subjects pleaded guilty. In December 2020, the third defendant was convicted of destruction of religious property, obstruction of religious beliefs, using a destructive device in relation to crimes of violence, possession of an unregistered destructive device, and conspiracy charges. The case is pending.	AGAAVE, specifically MVE
	An individual was arrested on state charges of attempted murder and criminal recklessness after allegedly barricading himself in his residence and shooting at law enforcement officers attempting to serve him an eviction notice.	AGAAVE, specifically SCVE
12 August 2017 Charlottesville, Virginia	An individual was arrested and federally charged with a hate crime resulting in death, hate crime acts involving attempt to kill, and bias-motivated interference with federally protected activity resulting in death on state charges for driving a vehicle into a crowd of protestors, killing one person and injuring at least 19 others. In December 2018, the subject pleaded guilty and was subsequently sentenced to life in prison.	RMVE
12 August 2017 Charlottesville, Virginia	An individual was arrested on state charges of discharging a firearm within one thousand feet of a school for shooting at a Black person at a protest. In May 2018, the subject pleaded no contest, and in August 2018, was sentenced to 8 years.	RMVE

(continued)

Date and location	Description	Category[b]
12 August 2017 Oklahoma City, Oklahoma	An individual was arrested and federally charged with malicious attempted destruction of a building used in and affecting interstate commerce by means of an explosive and attempted use of a weapon of mass destruction for plotting to detonate what he believed was a vehicle borne improvised explosive devise at a bank headquarters. In February 2019, the subject was found guilty. In March 2020, the subject was sentenced to 25 years.	AGAAVE, specifically MVE
18 August 2017 Kissimmee, Florida	An individual was arrested on state charges of premeditated murder for shooting and killing two law enforcement officers. In December 2019, the subject was convicted and was sentenced to death.	RMVE
20 August 2017 Livonia, Louisiana	An individual was arrested on a state charge of using, carrying, and/or possessing a firearm during a crime of violence and federally charged with carjacking for taking a hostage at gun point, discharging a firearm at the hostage and a convenience store clerk, robbing the convenience store, and attempting to rape two women. In September 2019, the subject pleaded guilty, and in February 2020, was sentenced to 20 years.	AGAAVE, specifically MVE
20 October 2017 Gainesville, Florida	Three individuals were arrested on state charges of attempted homicide for shooting at a group of protesters at a speaking event. In August 2018, one subject pleaded guilty to a state charge of accessory after the fact to attempted first degree murder, and in March 2019, was sentenced to 5 years. In February 2019, the second subject pleaded guilty to state charges of aggravated assault with a deadly weapon and possession of a firearm by a felon, and was sentenced to 15 years. In October 2019, the third subject pleaded guilty to assault and was sentenced to 5 years.	RMVE
3 December 2017 Somerset, Pennsylvania	An incarcerated individual allegedly attempted to kill a corrections officer because the officer was friendly with Black prisoners. The subject had an extensive criminal history and was serving multiple life sentences.	RMVE
2 January 2018 Orange County, California	An individual was arrested on state charges of murder with a hate crime enhancement for allegedly killing a homosexual Jewish college student.	RMVE
9 February 2018 Locust Grove, Georgia	An individual shot at three law enforcement officers, killing one and injuring two, who were attempting to arrest him. The subject died as a result of engagement with law enforcement.	AGAAVE, specifically SCVE
17 March 2018 Murfreesboro, Tennessee	An individual was arrested on state charges of homicide for killing his Black housemate by setting the victim on fire. In July 2019, the subject pleaded guilty and was sentenced to life.	RMVE
19 August 2018 Pittsburgh, Pennsylvania	An individual was arrested on state charges of homicide for allegedly stabbing and killing a Black person outside of a bar.	RMVE

(continued)

Date and location	Description	Category[b]
1 October 2018 Oklahoma	Law enforcement officers attempted to arrest an individual with a personalized violent ideology on state charges related to allegedly making online threats to kill law enforcement. During the arrest, the subject allegedly engaged in gunfire with law enforcement and was injured before being taken into custody.	DVE
23 October 2018 San Juan, Puerto Rico	Individuals were arrested for allegedly throwing rocks and other objects toward law enforcement officers and vandalizing property.	AGAAVE, specifically Puerto Rican National Violent Extremism
24 October 2018 Jeffersontown, Kentucky	An individual was arrested on state charges of murder, wanton endangerment, and attempted murder for allegedly shooting and killing two Black people in a grocery store. The subject was later federally charged with a hate crime resulting in death, use and discharge of a firearm to commit murder during and in relation to a crime of violence, and a hate crime with attempt to kill. In March 2021, the subject pleaded guilty to hate crimes and firearms offenses; the case is pending.	RMVE
27 October 2018 Pittsburgh, Pennsylvania	An individual was arrested and federally charged with obstruction of exercise of free religious beliefs resulting in death, use of a firearm to commit murder during and in relation to a crime of violence, obstruction of free exercise of religious beliefs resulting in bodily injury to a public safety officer, and use and discharge of a firearm during and in relation to a crime of violence for allegedly shooting and killing multiple people at a synagogue during a religious service. In January 2019, the subject was charged with additional hate crimes and firearms offenses. In total, 11 people were killed, and at least six others were injured. The case is pending.	RMVE
26 October 2018 Plantation, Florida	An individual with a personalized violent ideology was arrested and federally charged with use of weapons of mass destruction; transport or receive, in interstate or foreign commerce any explosive with the knowledge or intent that it will be used to kill, injure, or intimidate; transmit in interstate or foreign commerce any communication containing any threat to injure; knowingly deposits for mailing or delivery, or knowingly causes to be delivered by mail, anything declared non- mailable, with intent to kill or injure; uses/ carries an explosive to commit or during the commission of any felony, for mailing possible improvised explosive devices to US representatives, former US government officials, private citizens, and national media figures. In March 2019, the subject pleaded guilty, and in August 2019, was sentenced to 20 years.	DVE

(continued)

Date and location	Description	Category[b]
10 December 2018 Toledo, Ohio	Two individuals were arrested and federally charged for their involvement in planning terrorist attacks against a local bar and an interstate pipeline. The subjects pleaded guilty and in November and December 2019, they were sentenced to 15 years and 6 years, respectively.	AGAAVE, specifically AVE
18 January 2019 Islamberg, New York	Three individuals were arrested on state charge for plotting to attack Muslim residents of Islamberg, New York. In June 2019, one subject pleaded guilty to attempted illegal possession of a weapon, and in September 2019, was sentenced to 7 years. In June 2019, the other two subjects pleaded guilty to conspiracy in support of terrorism, and in August 2019, were sentenced to four to 12 years.	RMVE
15 February 2019 Silver Spring, Maryland	An individual was arrested and federally charged with unlawful possession of unregistered firearm silencers, unlawful possession of a firearm, and possession of a controlled substance. In October 2019, the subject pleaded guilty, and in January 2020, was sentenced to 160 months.	RMVE
25 March 2019 and 4 April 2019 New York and Missouri	Two individuals were arrested and federally charged with making a material false statement to the FBI in connection with their alleged plot to attack a mosque or synagogue. In July 2019, one subject pleaded guilty and was sentenced to time served. In August 2019, the other subject pleaded guilty, and in December 2019 was sentenced to probation.	RMVE
24 April 2019 San Diego, California	An individual was arrested and federally charged with malicious destruction of a building by means of fire for driving a vehicle into a building occupied by a cleared defense contractor and then setting the vehicle on fire. In June 2019, the subject pleaded guilty, and in November 2019, was sentenced to 7 years. The subject was also ordered to pay $93,633 to the victim.	AGAAVE, specifically AVE
27 April 2019 Poway, California	An individual was arrested on state charges of murder and attempted murder for allegedly conducting a shooting at a synagogue, killing one person and injuring three others. The subject was later federally charged with obstruction of free exercise of religious beliefs resulting in death and bodily injury, involving attempt to kill, use of a deadly weapon; hate crime acts, and damage to real religious property involved the use of a dangerous weapon or fire.	RMVE
6 June 2019 Brownsville, Texas	An individual was arrested on state charges of making terroristic threats on social media toward a synagogue, mosque, and a federal building. The subject was later federally charged with making a threat with explosive materials and false information and hoaxes. In August 2020, the subject pleaded guilty, and in November 2020, was sentenced to 2 years.	RMVE

(continued)

Date and location	Description	Category[b]
17 June 2019 Dallas, Texas	An individual shot at people and buildings, including a US Federal Courthouse. The subject died as a result of engagement with armed security officers.	AGAAVE
18 June 2019 Jacksonville, North Carolina	An individual was arrested for allegedly planning to conduct a mass shooting targeting synagogues.	RMVE
13 July 2019 Takoma, Washington	An individual threw incendiary devices at vehicles and outbuildings at a federal detention center and engaged responding law enforcement officers with an AR-style rifle. The subject died as a result of engagement with law enforcement.	AGAAVE, specifically AVE
28 July 2019 Gilroy, California	An individual with a personalized violent ideology attacked a festival, shooting and killing three people and injuring 16 others. The subject died as a result of engagement with law enforcement.	DVE
3 August 2019 El Paso, Texas	An individual was arrested on state charges of capital murder for allegedly attacking a retail store, shooting and killing 22 people and injuring 25 others.	RMVE
8 August 2019 Las Vegas, Nevada	An individual was arrested and charged federally with possession of an unregistered firearm or explosive device after discussing threats against multiple targets including a synagogue and an LGBTQ nightclub. In February 2020, the subject pleaded guilty, and in November 2020, was sentenced to 2 years.	RMVE
21 September 2019 Fort Riley, Kansas	An individual was arrested and federally charged with distributing explosives information and threatening interstate communication after providing bombmaking instructions online and conducting potential target selection. In February 2020, the subject pleaded guilty, and in August 2020, was sentenced to 30 months.	RMVE
10 December 2019 Jersey City, New Jersey	Two individuals shot and killed one law enforcement officer, and then attacked a kosher supermarket, shooting and killing three people and injuring three others, including two responding officers. The subjects died as a result of engagement with law enforcement.	RMVE
28 December 2019 Monsey, New York	An individual was arrested and charged federally with a hate crime act involving an attempt to kill, and obstruction of free exercise of religious beliefs involving an attempt to kill and use of a dangerous weapon, and resulting in bodily injury, for allegedly attacking the home of a Hasidic rabbi with a machete during a Hanukkah celebration, killing one person and injuring four others. The case is pending.	RMVE

[a]Unless otherwise noted, some of these matters are active/pending

[b]This column reflects FBI's categorization for purposes of this report and as required by statute. All defendants are presumed innocent until proven guilty beyond a reasonable doubt in a court of law

Federal Bureau of Investigation Department of Homeland Security; Strategic Intelligence Assessment and Data on Domestic Terrorism May 2021.

Appendix 4 (Source: USGS)

USGS Magnitude 2.5+ Earthquakes, Past Day August 19, 2021

3.3
6 km WSW of La Parguera, Puerto Rico
2021-08-19 11:35:27 (UTC-04:00)11.0 km
2.6
2 km E of Pāhala, Hawaii
2021-08-19 11:09:02 (UTC-04:00)36.3 km
3.3
6 km WSW of La Parguera, Puerto Rico
2021-08-19 10:17:36 (UTC-04:00)11.0 km
3.1
6 km E of La Parguera, Puerto Rico
2021-08-19 10:02:56 (UTC-04:00)11.0 km
3.1
Southeastern Alaska
2021-08-19 09:16:21 (UTC-04:00)10.0 km
3.0
5 km WSW of La Parguera, Puerto Rico
2021-08-19 07:56:26 (UTC-04:00)11.0 km
2.8
37 km SE of Mina, Nevada
2021-08-19 07:15:56 (UTC-04:00)8.1 km
2.9
47 km E of Juneau, Alaska
2021-08-19 06:11:43 (UTC-04:00)0.0 km
3.1
5 km SSW of Palmarejo, Puerto Rico
2021-08-19 05:17:45 (UTC-04:00)5.0 km
2.5
1 km NNW of Indios, Puerto Rico
2021-08-19 04:43:35 (UTC-04:00)6.0 km
2.6
31 km NNW of Mosquito Lake, Alaska
2021-08-19 03:29:51 (UTC-04:00)6.6 km
2.5
3 km S of Pāhala, Hawaii
2021-08-19 02:07:27 (UTC-04:00)32.2 km
2.6
53 km E of Juneau, Alaska
2021-08-19 02:01:25 (UTC-04:00)9.9 km
9 km ESE of Carson City, Nevada
2021-08-19 01:23:53 (UTC-04:00)11.3 km

2.6
Southeastern Alaska
2021-08-19 01:17:21 (UTC-04:00)3.7 km
2.6
Southeastern Alaska
2021-08-19 01:17:21 (UTC-04:00)3.7 km
3.4
18 km ESE of Ocotillo Wells, CA
2021-08-19 00:29:46 (UTC-04:00)3.0 km
2.9
80 km ENE of Chignik, Alaska
2021-08-18 22:42:46 (UTC-04:00)0.5 km
3.9
80 km ENE of Chignik, Alaska
2021-08-18 22:22:28 (UTC-04:00)12.2 km

2.8
18 km ESE of Ocotillo Wells, CA
2021-08-18 22:14:41 (UTC-04:00)3.3 km
2.9
133 km W of Gold Beach, Oregon
2021-08-18 21:35:19 (UTC-04:00)10.0 km
2.5
18 km ESE of Ocotillo Wells, CA
2021-08-18 18:40:26 (UTC-04:00)3.2 km
3.8
17 km ESE of Ocotillo Wells, CA
2021-08-18 17:50:22 (UTC-04:00)3.8 km
4.0
18 km ESE of Ocotillo Wells, CA
2021-08-18 17:45:14 (UTC-04:00)7.1 km
2.6
54 km SE of Perryville, Alaska
2021-08-18 17:39:56 (UTC-04:00)16.1 km
2.5
64 km NE of Chignik, Alaska
2021-08-18 17:21:53 (UTC-04:00)80.0 km
3.3
203 km SE of Perryville, Alaska
2021-08-18 16:06:53 (UTC-04:00)22.4 km
3.4
128 km SSE of Sand Point, Alaska
2021-08-18 15:30:57 (UTC-04:00)26.9 km
2.5
18 km ESE of Ocotillo Wells, CA
2021-08-18 14:34:26 (UTC-04:00)4.2 km
2.7
18 km ESE of Ocotillo Wells, CA
2021-08-18 14:34:03 (UTC-04:00)7.4 km
2.6
14 km N of Nikiski, Alaska
2021-08-18 12:24:30 (UTC-04:00)56.5 km

References

Akpan, N. (2016, October 26). America's HIV outbreak started in this city, 10 years before anyone noticed. *Science*.

Andrew, S. (2021, February 25). Traumatized and tired, nurses are quitting due to the pandemic. *CNN*.

Aydinalp, C., & Cresser, M. (2008). The effects of global change on agriculture. *American-Eurasian Journal of Agricultural & Environmental Sciences, 3*(5), 672–676.

Bambra, C., et al. (2020). The Covid-19 pandemic and health inequalities. *Journal of Epidemiology and Community Health*. https://doi.org/10.1136/jech-2020-214401

Barnard, A. (2011). *Social anthropology and human origins*. Cambridge University Press.

Beacham, A. M., et al. (2019). Vertical farming: A summary of approaches to growing skywards. *The Journal of Horticultural Science and Biotechnology, 94*, 277–283.

Bergin, P., et al. (2019) What is the threat to the United States? Terrorism in America after 9/11. *New America*.

Brown, R., Helmus, T., et al. (2021). *What do former extremists and their families say about radicalization and deradicalization in America?* RAND Corporation.

Busher, J. (2015) *What part of social networks play in radicalization?* (Paperback ed.). University of California Press.

Calkin, D., et al. (2014). PNAS (Proceedings of the National Academy of Science of the U.S.A).

CDC. (2000). General practice—Time for a new definition. *The BMJ*.

CDC (Centers for Disease Control and Protection). National Environment Public Health Tracking.

Climate Central. (2021). Fourth national climate assessment: Chapter 14. In *Human health, climate change and health*.

Congress. (2001). Section 802 of the USA PATRIOT Act [Pub. L. No. 107-52].

Congressional Research Service. (2021).

Cook, T. (2015, April 3). Ground-shaking research: How humans trigger earthquakes. *EARTH*.

Crosby, A. (2003). *America's forgotten pandemic: The influenza of 1918*. Cambridge University Press.

Cutter, S. L. (2001). Charting a course for the next two decades. In S. L. Cutter (Ed.), *American hazardscapes: The regionalization of hazards and disasters*.

Daley, D. (2020, September 27). *I cry for the mountains*. Enterprise-Record.

Davies, E. BBC ScienceFocus. *Everyday Science*.

Desert Research Institute. (2021, May 25). *Does cold wildfire smoke contribute to water repellent soils in burned areas?*

Duong, F. (2021). Dinosaur-killing asteroid probably created the Amazon rainforest. *SCIENCE*.

Ebarts, B., & Stein, G. (2020). *U.S. Fire Department Profile 2018*.

P. A. Erickson, *Effective Environmental Emergency Responses*,
https://doi.org/10.1007/978-3-031-05893-6

Ellis, J. J. (2021a). *The cause: The American revolution and it's discontents, 1773–1783.* W. W. Norton & Company.

Ellis, J. J. (2021b). *The cause: The American revolution and it's discontents, 1773–1783.* W. W. Norton & Company. (Original source: An Address to the People of Great Britain, American Archives).

Enikolopov, R., & Petrova, M. (2017). *Mass media and its influence on behaviour.* CREi (Centre de Recerca en Economia Internacional).

EPA. (2020). *The legacy of EPA's acid rain research.*

EPA. (2021a). *Climate change indicators.*

EPA. (2021b). *Sources of greenhouse gas emissions.*

Erickson, P. A. (1979). *Environmental impact assessment: Principles and applications.* Academic.

Erickson, P. A. (1994). *A practical guide to environmental impact assessment.* Academic.

Erickson, P. A. (1999). *Emergency response planning for corporate and municipal managers.* Academic.

Erickson, P. A. (2006). *Emergency response planning for corporate and municipal managers* (2nd ed.). Elsevier.

FEMA. (2017, June). *National earthquake loss estimate.*

Fernandez-Armesto, F. (2001). *Civilizations: Culture, ambition, and the transformation of nature.* The Free Press.

Finkelhor, D., et al. (2009). *Children's exposure to violence: A comprehensive national survey.* DOJ, Office of Justice Programs, Office of Juvenile Justice and Delinquency Prevention (Updated 2017).

Folger, T. (2021). *The next green revolution.* National Geographic.

Frohnen, B. (2001). Tradition, habit, and social interaction: A response to Mark Bevir. *Humanitas, 14*(1), 108–116.

Gavrilets, S., et al. (2016). Convergence to consensus in heterogeneous groups and the emergence of informal leadership. *Scientific Reports, 6*, 29704. https://doi.org/10.1038/srep29704

Gibbons, S. (2017, October 2). *How humans are causing deadly earthquakes.* National Geographic.

Gray, P. (2011). *How hunter-gatherers maintained their egalitarian ways.* Psychology Today.

Guary, D., & Davis, S. (2020). *Full cost of California's wildfires to the U.S. revealed, nature sustainability.* University of California, Irvine.

Heldman, C. (2011, August 29). *Hurricane Katrina and the demographics of death.* Sociological Images.

Horton, R. (2020). Covid-19 is not a pandemic. *Lancet, 396*, 874.

Insurance Information Institute. (2021). *Annual number of acres burned in wildland fires, 1980–2020.*

International Association of Fire Chiefs. (2018). NFPA 1582.

IPCC. (2021). *Intergovernmental Panel on Climate Change, IPCC Press Release: Climate change widespread, rapid, and intensifying.*

IRG (Integral Risk Global). *Multi person helicopter rescue.*

Jones, S., et al. (2020). *Domestic terrorism in the United States.* Center for Strategic and International Studies (CSIS).

Kassinove, H. (2012). *How to recognize and deal with anger.* American Psychological Association.

Lindsey, R. (2020–2021). *Climate change: Global sea level.* Climate.gov. science & information for a climate-smart nation.

Madison, J. (1887). Federalist No. 10. In *The Federalist Papers.* Country House Publishing (2015).

McGinty, E., et al. (2020). Psychological distress and loneliness reported in 2018 and April 2020. *Journal of the American Medical Association.* https://doi.org/10.1001/jama.2020.9740

McNeill, W. (1977, 1989, 1998). *Plagues and peoples.* Random House/Anchor Books.

McQuail, D. (1979). The influence and effects of mass media, mass communication and society. .

Metulka, R. (2014). *The history of the electric car.* U.S. Department of Energy.

Meyer, R. (2021). Why America doesn't really make solar panels anymore. *The Atlantic.*

MHA. (2021). Mental Health America, Inc. Black and African American Communities and Mental Health.

Milani, L. (2017). *The process of radicalization*. Futures Without Violence.

Miller, E. (2014). *Terrorist attacks in the U.S. between 1970 and 2013: Data from the Global Terrorism Database (GTD)*. START (National Consortium for the Study of Terrorism and Responses to Terrorism).

Moseley, C. (2013). *The economics of large fires*. Digital Commons as University of Nebraska-Lincoln.

National Center for Environmental Health. (2020). *Diseases carried by vectors*.

National Equity Atlas. (2021). *Percent of households without vehicle, USA, 1980–2117*.

News: Medical-Life Sciences. (2006, October 23). *1,500 animal species practice homosexuality*.

Nolon, J. R. (2006, September 8). *A diagnostic approach to evaluating governmental land use control*. Pace University, Pace Law Faculty Publication (Article 8).

Padhy, K. (2020, December). U.S. eviction crisis. *BBC*.

Pennisi, E. (2020). Humans have altered North America's ecosystems more than melting glaciers. *SCIENCE*.

Pescaroli, G., & Alexander, D. (2015). A definition of cascading disaster and cascading effects: Going beyond the "toppling dominos" metaphor. *Global Risk Forum (GRF), Planet@Risk, 3*(1).

Peter G. Peterson Foundation. (2021). *Note from Michael Peterson on the impact of the pandemic: How did Americans spend their stimulus checks and how did it affect the economy*. Policy Advice, 2021, Earthquake Statistics by Year.

PRB (Population Reference Bureau). (2002). *Children's environmental health: Risks and remedies*.

Reardon, S. (2021, August 19). The true Haiti earthquake death toll is much worse than early official counts. *Scientific American*.

Reid, K. (2021, August 23). *Haiti earthquake: Facts, FAQs, and how to help*. World Vision.

Rohr, J., et al. (2011). *Frontiers in climate change-disease research*. Trends in Ecology &Evolution-Cell Press.

Roman, J., et al. (2020, November–December). NFPA (National Fire Protection Association).

Rutherford, A. (2019). *Humanimal: How Homo sapiens became nature's most paradoxical creature*. The Experiment, LLC.

Schuur, E. A. G., et al. (2013). *Expert assessment of vulnerability of permafrost carbon to climate change*. Springer.

Schwimmer, B. (2001). *Hawaiian Kin terms*. University of Manitoba (Updated: August 24, 2011).

Science News. (2022a, February 4). *Preventing pandemics costs far less than controlling them*. Duke University.

Science News. (2022b). *Climate change has likely begun to suffocate the word's fisheries*. American Geophysical Union.

Select Bipartisan Committee. (2006). *A failure of initiative: Final report*. U.S. Government Printing Office.

Shute, N. (2021, January 24). When data shed light on societal challenges. *Science News (editorial)*.

Simon, R. (2021, April 16). Covid 19's toll on U.S. business: 200,000 extra closures in pandemic's first year. *The Wall Street Journal*.

Simpson, D. W., et al. (1981). *The 1906 San Francisco earthquake and the seismic cycle, earthquake prediction: An international review* (Maurice Ewing series 4, pp. 126–140). American Geophysical Union.

Soh, C. H. (2000). *Director general: Fire and rescue services, Malaysia (retired)*. From: Lectures in Vietnam, Thailand, Malaysia, Singapore, Indonesia, and Fiji.

Subedi, D. (2017). Early warning and response for preventing radicalization and violent extremism. *Peace Review, 29*(2), 135–143.

Summer, A. (2019). *Social media's growing impact on our lives*. American Psychological Association.

Sundaram, A. (2020). Yelp data shows 60% of business closures due to the Coronavirus pandemic are now permanent. *CNBC*.

Tattersall, I., & Desalle, R. (2019). *The accidental Homo Sapiens*. Pegasus Books.

Taylor, A. (2001). *American colonies*. Viking Press.

The Guardian. (2011). *What are the main man-made greenhouse gases*.

The Guardian. (2021). *Scientists watch giant 'doomsday' glacier in Antarctica with concern*.

U.S. Census Bureau. (2019). Release number CB19-TPS.51.

U.S. EPA (Environmental Protection Agency). *A closer look: Land loss along the Atlantic Coast*.

U.S. Fire Administration. (2016). *Critical health and safety issues in the Volunteer Fire Service*.

U.S. Fire Administration. (2021). *Fire estimate summary, April 2021: Residential building cooking fire trends*.

U.S. Government Printing Office. (2006a). *Federal response to Hurricane Katrina: Lessons learned*.

U.S. Government Printing Office. (2006b). *A failure of initiative: Final report of the select Bipartisan Committee to investigate the preparation for and response to Hurricane Katrina*.

University of Leeds. (2021). Himalayan glaciers melting at 'exceptional rate'. *Science News*.

University of Michigan. (2017). Center for Sustainable Systems (CSS). U.S. Cities Factsheet.

University of Michigan. (2019). Center for Sustainable Systems (CSS). U.S. Cities Factsheet.

USGS (United States Geological Survey). (2014). *How would sea level change if all glaciers melted*.

World Meteorological Organization. (2021). *WMO recognizes a new Arctic temperature record*.

York University. (2021, October 21). Northern lakes warming six times faster in the past 25 years. *ScienceDaily*.

Zaraska, M. (2021). *The tomatoes at the forefront of a food revolution*. British Broadcasting Corporation.